D0192946

Richard Feynman

1918 – 1988

BY THE SAME AUTHORS

Being Human
Time and Space
Fire on Earth
Watching the Weather

THE 'IN 90 MINUTES' SERIES

Halley
Einstein
Darwin
Curie

JOHN AND MARY GRIBBIN

Richard Feynman

A LIFE IN SCIENCE

A DUTTON BOOK

DUTTON
Published by the Penguin Group
Penguin Books USA Inc., 375 Hudson Street,
New York, New York 10014, U.S.A.
Penguin Books Ltd, 27 Wrights Lane, London W8 5TZ, England
Penguin Books Australia Ltd, Ringwood, Victoria, Australia
Penguin Books Canada Ltd, 10 Alcorn Avenue,
Toronto, Ontario, Canada M4V 3B2
Penguin Books (N.Z.) Ltd, 182–190 Wairau Road, Auckland 10, New Zealand

Penguin Books Ltd, Registered Offices: Harmondsworth, Middlesex, England

Published by Dutton, an imprint of Dutton Signet,
a division of Penguin Books USA Inc.
First published in Great Britain by Penguin Books Ltd.

First Dutton Printing, July, 1997
10 9 8 7 6 5 4 3 2 1

CIP data is available.

ISBN: 0-525-94124-X

Printed in the United States of America

This book is printed on acid-free paper. ♾

For Jacqueline Shaw,
Richard Feynman's sister-in-law,
who put the idea into our minds

I wonder why. I wonder why.
I wonder why I wonder.
I wonder *why* I wonder why
I wonder why I wonder!

RICHARD FEYNMAN

Contents

Acknowledgements *xi*
Prologue: 'We love you Dick' *xiii*

1 A fascination with physics *1*
2 Physics before Feynman *25*
3 College boy *45*
4 Early works *70*
5 From Los Alamos to Cornell *91*
6 The masterwork *120*
7 The legend of Richard Feynman *138*
8 Supercool science *155*
9 Fame and (some) fortune *173*
10 Beyond the Nobel Prize *189*
11 Father figure *204*
12 The last challenge *222*
13 The final years *241*
14 Physics after Feynman *260*

Epilogue: In search of Feynman's van *281*
Bibliography *285*
Index *288*

Acknowledgements

Many people gave up time to talk to us about their personal and professional memories of Richard Feynman. Others took the trouble to answer specific queries by mail, e-mail or telephone. Without them, this book could not have been an accurate portrayal of the best-loved scientist of our times. We thank especially Joan Feynman, Carl Feynman, Michelle Feynman and Jacqueline Shaw from Feynman's immediate family circle; James Bjorken, Norman Dombey, David Goodstein, James Hartle, Robert Jastrow, Daniel Kevles, Hagen Kleinert, Igor Novikov, Kip Thorne and Nick Watkins from the world of physics; Feynman's former secretary, Helen Tuck; and Ralph Leighton, who knew Feynman as well as anybody did in the last decade of his life. Even where these people have not been quoted directly, their contributions have helped to shape the image of Richard Feynman in our minds, which we hope comes through in this book.

We have also drawn on published accounts of Feynman's life and work, cited in the text and referred to in full in the Bibliography. Where possible, we have checked important stories about Feynman with their sources; but, of course, we have had to rely on the secondary sources in cases where the originators of the stories are no longer alive, or were otherwise unavailable.

Michael Shermer went to enormous trouble to arrange many interviews for us on a visit to Caltech, and Jagdish Mehra, who was the last person to interview Feynman formally about his life and work, gave permission for us to quote from his own book *The Beat of a Different Drum*, which remains the definitive technical account of the life and science of Richard Feynman, at a more academic level than the present book.

Benjamin Gribbin spent many hours transcribing recordings of interviews with scrupulous accuracy and unfailing good humour, and Jonathan Gribbin prepared the diagrams with speed and skill. Christopher Allen carried out the picture research, and the archivists at Princeton University and Caltech, respectively Ben Primer and Charlotte Erwin, helped us to find source material, as did Karl Berkelman at Cornell University, Helen Samuels at MIT, and Roger Meade at the Los Alamos National Laboratory.

Prologue: 'We love you Dick'

Does the world really need *another* book about Richard Feynman? We think so, or we wouldn't have written it. And this is why. Richard Feynman was the best-loved scientist of modern times, perhaps of all times, and that is something that simply does not come across in any of the other books about the man and his work. There have been books about Feynman the character, a wise-cracking entertainer who imparted not a little worldly wisdom along with his anecdotes; there have been books about Feynman the scientist, putting his work in the perspective of physics in the second half of the twentieth century; there has even been a picture book, combining the illustrations with reminiscences about Feynman by his family and friends. But nobody has captured the essence of Feynman's science and the essence of Feynman's persona in one book. This is especially odd because, of all the scientists of modern times, Feynman seems to have been the one who had the best 'feel' for science, who understood physics not simply in terms of lines of equations written on a blackboard, but in some deep, inner sense which enabled him to see to the heart of the subject.

This doesn't mean that Feynman lived his life 'like a scientist', in the stereotypical sense of being a cold-blooded logician in everyday life. Far from it. The point is that he did physics 'like a human being', carrying into the world of science his inbuilt sense of fun, his irreverance, and his liking of adventure and the unexpected. The way Feynman did his physics depended on the kind of person he was, far more than in the case of any other physicist we know. It is impossible to understand Feynman's science properly without understanding what kind of a person he was, and nobody put more life into science than he did.

Equally, it is impossible to understand what kind of man Feynman was without understanding at least something of the science that was so important to him. A fun-loving, adventurous character like Feynman was attracted to physics because physics is fun, and offers opportunity for adventure. You may find that hard to believe. But what's wrong with the public image of physics is not so much the science itself as the way that the science is taught and portrayed. Perhaps Feynman's greatest achievement was as a teacher, conveying the fun of science, and entertainer, providing an image of science that cut right across the stereotypes. Ralph Leighton describes Feynman as a 'shaman of physics'. Feynman talked of nature as 'She' or 'Her', and seemed to have a contact with the way the world works that few people have. When he gave lectures, he brought his audience into contact with nature in ways that they could not achieve on their own, allowing them to see nature differently, in a transforming experience, so much so that often when he explained some subtle point in a way that they could understand the audience would break out into spontaneous applause, even laughter. The physicist Freeman Dyson has commented,[1] 'I never saw him give a lecture that did not make the audience laugh', but the laughter stemmed as much from the pleasure of finding things out as from the jokes that Feynman cracked.

After this experience, people would often have a memory of understanding something, but couldn't always quite reconstruct how it was they had understood – Feynman would raise people to a level of understanding that they had never before achieved, but then they couldn't quite remember how he had done it. Even fellow scientists sometimes felt this way about a Feynman lecture – Leighton recalls his own father, one of Feynman's colleagues at Caltech, remarking on this almost transcendental experience. People who attended Feynman's lectures say that they seemed like magic, almost literally spellbinding, while people who met him report the same sort of feeling, an awareness of being in the presence of something special, even when they can't quite put their finger on why. They just felt changed by the experience. And people who never met Feynman still write to Leighton to say that they have been inspired by Feynman's example. It may well be that he will be remembered more in this way, as a 'wise man', rather than for the specific aspects of the science that he was involved with.

This would be appropriate, and perhaps what Feynman himself would have wanted. To Feynman, love was more important than science; but it just happened that, as well as loving people, he loved physics.

And people, including physicists, loved him. In an obituary published in *Nature* on 14 April 1988 (volume 332, page 588), Hans Bethe, who had been Feynman's boss both at Los Alamos and at Cornell, said 'more than other scientists, he was loved by his colleagues and his students'. The day Feynman died, the students at Caltech hung a banner across the 11-storey library building on the campus. The message on the banner read: 'WE LOVE YOU DICK'. Around the world, many people who hadn't even met Feynman felt a sense of personal loss when he died. Neither of us ever met him; but the physicist half of the partnership (JG) was exactly the right age to be among the first undergraduates to benefit from Feynman's *Lectures on Physics* while at university. The clarity of those lectures helped to shape his career, and reinforced his own feeling that science, even at research level, could still be fun. Reading books and papers by Feynman over the years, and seeing him on TV, reinforced that belief, and made Feynman seem like an old friend.

But to many people who felt the same way, Feynman was, more than any other great scientist of modern times, 'famous for being famous'. The name of Stephen Hawking is inextricably linked with black holes; Albert Einstein's with relativity theory; Charles Darwin's with evolution. But Feynman? To many non-scientists, he was just 'a scientist'. This is ironic, because Feynman's greatest work was actually in the area of quantum theory, a subject of enormous fascination to non-scientists today. We want to explain why this work was so important, and how it lies at the heart of investigations of the quantum mysteries today; but we also want to share with you our understanding of the kind of man who carried out that work.

Even today, writing seven years after Feynman died in 1988, it is far too soon to produce a definitive account of the historical importance of the man and his work. We don't claim that this is more than a personal view of our subject, but it is one we have arrived at through a long (if one-sided) association with his works, and through recent discussions with Feynman's family and friends.

The one thing that is clear above all else in Feynman's character, from his own work and from conversations with people who knew him, is passion. His passion for physics, for drawing, for drumming, for life itself and for his jokes. Of course Feynman's own anecdotes, gathered together by Ralph Leighton and published in two volumes, tend to portray Feynman as a larger than life, legendary scientific superman and scourge of established authority. Were those stories accurate? We asked Feynman's sister, Joan, on a visit to Pasadena in April 1995. 'It's easy to tell which stories are accurate', she replied. 'How?', we asked. 'My brother didn't lie.'

Ralph Leighton, to whom the stories were told, agrees, but stresses that Feynman was a showman, who loved telling stories.[2] The stories were all true, in that they were about real things that had happened to Feynman; but he used to try telling them in different ways, with different emphasis, until he found the way that worked best. They were not, after all, just anecdotes; in many cases, the stories became parables, and have a moral, telling you something about the right way to live and how to get on in the world, as well as offering amusement and entertainment.

There is indeed a legend growing up around Richard Feynman; but there is truth behind the legend.[3] In the classic western *The Man Who Shot Liberty Valence*, a reporter is faced with a choice between printing the truth about the early career of a great man, or the legend, and in a memorable moment decides to 'print the legend'. We don't intend to go that far, although we agree with the spirit of that decision. We offer you something of the legend of Richard Feynman, but also something of the man behind the legend; and we hope we can put across the importance of his scientific work in language that non-scientists can both understand and enjoy. That, after all, is what Feynman himself would have wanted.

John Gribbin*
Mary Gribbin
March 1996

*http://star-www.maps.susx.ac.uk/people/jrgribbin.html

Notes

1. See Freeman Dyson *From Eros to Gaia* (Pantheon, New York, 1992).
2. Joan Feynman, interviewed by JG in April 1995, said that according to her mother 'when Richard was very little he couldn't decide whether he wanted to be a comedian or a scientist, so he combined the two options'.
3. Interviewed by JG in April 1995, David Goodstein, who is Professor of Physics and Vice Provost at Caltech, said, 'Feynman is a person of historic proportions; he deserves the kind of attention that he's gotten, in my opinion.'

— 1 —

A fascination with physics

Family legend has it that when his wife Lucille became pregnant for the first time, Melville Feynman commented 'if it's a boy, he'll be a scientist'.[1] The baby was born on 11 May 1918 in Manhattan, and brought up in Far Rockaway, New York; he was named Richard Phillips Feynman,* and he grew up to be the greatest scientist of his generation. He not only won the Nobel Prize for Physics for his first major contribution to science, but carried out at least two other pieces of research that were worthy of the prize; he was one of the leaders of the team that worked on the Manhattan Project, to develop the atomic bomb; and he was, above all, a great teacher who encouraged generations of students to think about physics in a new way.

Melville Feynman has to take some of the credit for this, because he deliberately set out to stimulate his son to think, from an early age, in a 'scientific' way. When the boy was sitting in his high chair, Melville would play games with him using a collection of coloured bathroom tiles. At first, the game mainly involved setting up a row of tiles on end, in any order, and toppling them, like dominoes; but soon they moved on to setting up patterns, maybe two white tiles followed by a blue one, then two more white and another blue, and so on. The young Feynman – called Ritty or Richy by his parents, family and friends – became very good at the game, which his father had started in a conscious attempt to get young Ritty to think about patterns and the basics of mathematical relations.[2]

Melville encouraged his son's interest in science in the obvious ways – buying a set of the *Encyclopaedia Britannica*, taking Ritty on

*The name is pronounced, rather appropriately, 'Fine man'.

trips to the American Museum of Natural History, and so on. But even the conventional sources of information were used by Melville as jumping-off points for extrapolations which made the dry material come alive, and which brought home to Richard the magical, mysterious aspects of science. When the *Britannica* mentioned that a long-extinct dinosaur had been 'twenty-five feet high' and had a head 'six feet across', Melville would stop reading and explain what that meant – that if the dinosaur stood in the front yard of the house in Far Rockaway, he would be able to look in through the second-floor window, but his head would be too big to fit through the window.

But the special nature of Richard's relationship with his father, and the special nature of the way in which Melville encouraged the younger Feynman's fascination with science, is highlighted by two of Richard Feynman's favourite anecdotes about his father.

The first dates back to summers spent in the Catskill Mountains, where families from New York would go to escape the heat of the city. Mothers and children would stay in the mountains for several weeks, but the fathers of the families still had to work in the city, only visiting their families at weekends. On long weekend walks in the woods, Melville introduced Richard to many of the wonders of nature – but with his typical sideways manner of looking at the world. So when one of the other children pointed out a bird to Richard and asked if he knew its name, he had to reply that he didn't. Triumphantly, the other kid named the bird, sneering that 'your father doesn't teach you anything'. 'But', Feynman tells us,[3] 'it was the opposite.' His father had already pointed out that kind of bird:

> 'See that bird?' he says. 'It's a Spencer's warbler.' (I knew he didn't know the real name.) 'Well, in Italian, it's a *Chutto Lapittida*. In Portuguese, it's a *Bom da Peida*. In Chinese it's a *Chung-long-tah*, and in Japanese it's a *Katano Tekeda*. You can know the name of that bird in all the languages of the world, but when you're finished, you'll know absolutely nothing whatever about the bird. You'll only know about humans in different places, and what they call the bird. So let's look at the bird and see what it's *doing* – that's what counts.'

So Richard learned, at a very early age, the difference between knowing the name of something and knowing something. To such a person, it made perfect sense, years later when he was in graduate college, to ask a baffled librarian where he could find 'the map of a cat', and to be equally baffled by her reaction to this simple request. The actual telling of this story, many years later, also gave a fundamental insight into Feynman's childhood and upbringing. While going through that story with Feynman, not long before Feynman died, Ralph Leighton said to him, 'there's all this about your father, but what did your mother teach you?' He replied, 'My mother taught me that the highest forms of understanding that we can achieve are laughter, and human compassion.'[4]

The second key anecdote from Richard's early childhood concerns the occasion when he noticed the odd behaviour of a ball left lying in his little wagon when he pulled the wagon forward. The ball rolled to the back of the wagon, then, when the wagon stopped the ball rolled to the front. He asked his father why this happened, and got this reply:

> That, nobody knows. The general principle is that things which are moving tend to keep on moving, and things which are standing still tend to stand still, unless you push them hard. This tendency is called 'inertia', but nobody knows why it's true.

This represents a deep insight into the nature of physics and the nature of the world, and it was examples like this that encouraged Richard Feynman, in later years, to question everything, to search for underlying truths, and never to believe that just because some process had been labelled meant that it was understood.*

But there is another aspect to this way Melville had of teaching his son, which has echoes in the way Feynman later used his own anecdotes to bring out highlights of his own life when he became a storyteller in his turn. The stories don't have to be literally 'true', in every detail, in order to make a valid point. As Feynman himself said, he knew full well that the bird being described by Melville wasn't really called a 'Spencer's warbler', and that the foreign 'names' his father made up for the bird were just nonsense words.

*Intriguingly, one of Feynman's own insights into the nature of the world now provides us (although it was not appreciated in his lifetime) with one way of explaining what inertia 'really is'; see Chapter 14.

But he also knew that that didn't matter – that, indeed, the whole point of this particular story was that names didn't matter, so if Melville wanted to call the bird a Spencer's warbler he was fully entitled to do so. Richard Feynman's own stories should always be understood in this spirit – that as long as the underlying message is correct, the details and emphasis can be adjusted to improve the impact of the story. Joan Feynman's brother didn't lie, but as a great showman he presented his stories in the best possible light. As he said of his father's stories, 'I knew that they weren't quite accurate, and yet they were utterly accurate, if you see what I mean, in the character of the story he was trying to tell me.'[5] We could say the same about his own stories, especially when, for example, he quotes childhood conversations with his father verbatim, as if he had total recall, when in fact he was making up dialogue to match what he remembered of the occasion. The truth in Richard Feynman's anecdotes is a much deeper truth than the trivia of exactly what words were said on a particular day in the 1920s.

But if Richard learned so much about how to think about science and the world – not just an accumulation of scientific facts – from his father, where did Melville learn to think about the world in this way? Melville's own father, Richard's grandfather, was, apparently, also interested in mathematical and scientific ideas, so to that extent, at least, there was a tradition of science in the family. This offers hope for all of us; even if we cannot aspire to being a Richard Feynman, at least we can aspire to being a Melville Feynman – to have an understanding and enthusiasm for nature, and to pass that enthusiasm on to a child, even without the detailed mathematical knowledge that a professional scientist needs. But neither Richard's father nor his grandfather had an opportunity to develop their interest into a career.

Melville had been born in 1890. He was the son of Jakob and Anne Feynman, Lithuanian Jews who lived for a time in Minsk, in Byelorussia, and emigrated to the United States in 1895. The family settled in Patchogue, on Long Island, and Melville was initially taught at home, by his father (a precursor of his own relationship with Richard), but later attended the local high school. He wanted to become a doctor, but there was no way the family could afford to support the education required to fulfil his ambition, so instead he enrolled in a college to study homoeopathic medicine. Even these studies proved impossible to sustain financially, and Melville dropped out of college and into a variety of occupations, at none of which he was particularly successful, although he always managed to keep the family afloat, even through the Depression. He finally settled in the uniform business, providing ample opportunity for Richard to learn at first hand the difference between formal authority represented by a uniform and the frail human being inside the uniform. On one occasion, Feynman recalled, his father showed him a picture in the newspaper of the Pope, with people bowing down in front of him. 'What's the difference', Melville asked Richard, 'between this man and all the others?' He immediately answered his own question. 'The difference is the hat he's wearing. But this man has the same problems as everybody else: he eats dinner; he goes to the bathroom. He's a human being.'[6]

The parents of Lucille Phillips, Richard Feynman's mother, both came to the United States as young children. Her maternal grandfather (Richard's great-grandfather) was a Polish Jew who was involved in anti-Russian activities in the 1860s and 1870s, was imprisoned and sentenced to death, but escaped and eventually made his way to America, where his children later joined him. The eldest daughter among those children, Johanna Helinsky, worked with her father in the watchmaking store he opened on the Lower East Side in New York, and it was there that she met her future husband, Richard Feynman's maternal grandfather.

Henry Phillips was born in Poland, but lost his parents as a child and spent some time in an English orphanage, where he was given his name, before being sent on to America to seek his fortune. Unlike many immigrants in a similar position, Henry Phillips really did succeed in making a modest fortune. He started out selling needles and thread door-to-door from a pack on his back, and went on, with Johanna, to develop a successful millinery business, which thrived until changing fashions at the end of the First World War saw the hat business go into decline. Henry met Johanna when he had a watch that needed repairing, and took it into a watchmaking store where he was surprised to find the job being done by a beautiful young woman. They soon married, went into business together, and during the height of their success in the hat trade they moved to the Upper East Side, on 92nd Street, where Lucille Phillips (the youngest of five children) was born in 1895.[7] The family later moved to a large house with a big garden in Far Rockaway, which was then a semi-rural community in Queens County, at the southern tip of Long Island.

As the daughter of a successful businessman, Lucille was educated at the Ethical Culture Institute (where she was followed, nine years later, by Robert Oppenheimer), and intended to become a kindergarten teacher. But just after she graduated from high school, when she was 18 years old, she met Melville Feynman; they hit it off at once, and almost immediately he asked her to marry him. Her father wouldn't give his permission for her to marry so young, so they had to wait until 1917, after she had turned 21. At first, the newly married couple lived in upper Manhattan; Richard Phillips Feynman was born there, in a Manhattan hospital, a year after their marriage.

If Melville Feynman contributed, at least in part, to his son's becoming a scientist, Lucille had an equally great influence on him through her sense of humour, warmth and compassion. Joan Feynman feels that the role of their mother has been downplayed in most versions of the Feynman legend, leaving her in the shadows of the father who turned young Ritty on to science. Perhaps that is understandable, at least from the point of view of those recounting the legend. After all, many of us have mothers who have a wonderful sense of humour and are full of compassion, but very few people have fathers like Melville Feynman, so his part in the story seems at first sight more interesting and more profound. Without Lucille's influence, though, Richard Feynman might well have become a more or less conventional, dry as dust academic, rather than the safecracking, bongo-playing figure of legend. It is, after all, the combination of serious science, a sense of fun and the very sane view that 'the highest forms of understanding we can achieve are laughter and human compassion'[8] that made Feynman so special, and that combination is found in neither of his parents alone, but in both of them put together. And if any further proof of Lucille's influence on her son were needed, she was a great storyteller. Joan recalls:

> wonderful memories of evenings at the supper table when Richard was home from college and he and Mother would get going. My father and I would laugh so hard that our stomachs hurt and we would beg for mercy, but they wouldn't stop until I had fallen off my chair and was literally rolling on the floor.[9]

Even Lucille's good humour and compassion were severely tested, however, early in 1924, when Richard was five. She had another son, Henry Phillips Feynman, who was born on 24 January that year, but lived only for a month and a day, dying on 25 February. It wasn't until Richard was nine that his sister Joan was born; but that doesn't mean that he led anything like the usual life of an 'only child' for the first nine years of his life.

The Feynman family moved a couple of times when he was very small, but settled in Far Rockaway, where they shared Lucille's father's house with her sister Pearl and her family. That family included a son Robert, three years older than Richard, and a daughter Frances, three years younger than him. So he was in the middle of an extended family of children that were in fact cousins, but lived like siblings. The reason for the house-sharing was financial. Pearl's husband, Ralph Lewine, worked in the shirt business, but never achieved as much success as Melville did in his own line of business. The Feynman family was far from poor; they weren't as well off as Lucille's parents had been, but Joan Feynman recalls that they were always comfortable financially, right through the Depression years. Living in such close proximity wasn't always easy, at least for the adults in the two families (and, of course, the very fact that the house had been passed on by Henry Phillips was a constant reminder to both Melville and Ralph that they had not achieved as much as their father-in-law), and shortly after Joan was born, when Richard was 10, the Feynman family moved out to the nearby town of Cedarhurst. But within a couple of years they had returned, and although neither son-in-law ever became as successful in business as Henry Phillips, thanks in part to the house they had inherited from him both families survived the Depression in relative comfort. Joan remembers that she 'had nice clothes from good stores in New York', and that there was a woman who came in every day to clean and do laundry. 'Before the war, we had a new car every year (usually an Oldsmobile).'[10]

Even Melville, usually so iconoclastic and unwilling to be bound by convention, had one blind spot, though. True to his word, he encouraged Richard to take an interest in science. But he never attempted to rouse any similar enthusiasm in Joan. In the 1930s, it was almost inconceivable, even to someone as broadminded as Melville Feynman, that a girl could become a scientist. But Joan became a scientist anyway, ending up in space research at the prestigious Jet Propulsion Laboratory in Pasadena – she became, in fact, exactly the kind of scientist that Melville must have imagined Ritty might become. It all started when she would hear Melville and Richard talking about all these interesting things, and later she would ask her brother about what she had overheard. Soon, he was explaining things to her in the same way that he had learned them from their father, becoming a scientific raconteur (albeit to an audience of one) in his early teens.[11] Joan, too, helped to influence her brother's development, and likes to describe herself as 'Richard Feynman's first student'.[12]

It started when she was still a baby, and Richard had the duty of looking after her. Propped up in her baby carriage, she would watch Richard and a friend tinkering with the collection of wires, batteries and other electrical bits and pieces that they called their 'laboratory'. The family had a dog at the time, which had been taught tricks, and Richard reasoned that since his sister was brighter than the dog, she ought to be able to do better tricks. He decided to teach her arithmetic, in order to impress his friends, and encouraged her to learn by allowing her to pull his hair if she got the sum right. Joan still recalls standing in her crib, at the age of about three, 'yanking on his hair with great delight' having just learned to add two and three.

As Joan got bigger, so did her tasks. At five, she was a paid lab assistant, earning 2 cents a week for carrying out odd jobs and sometimes playing the part of the magician's assistant, sticking her finger in a small spark gap and enduring a modest electric shock, again to amaze Richard's friends. No anecdote sums up their relationship better; the hero-worshipping younger sister knew that her big brother would never hurt her, and trusted him to keep the shock at the level of mild discomfort, even though the sparks that leapt across the gap when no finger was in place looked terrifying to anyone not in the know. In exchange, as well as the financial rewards, Richard introduced her to the wonders of the world, showing her the stars and demonstrating centrifugal force by whirling a glass of water in an upside down arc without spilling a drop (except on one memorable occasion when the glass slipped out of his hand and flew across the room).

One of the things Richard showed her has stayed vividly in Joan's mind. She recalls that the household was run in a very orderly fashion, with strict rules about things like bedtime. As the youngest child in the household, she went to bed first. But one night, when she was about four years old, her brother, then about 13, got permission to wake her up. He told her he had something wonderful to show her, and took her out into the middle of a nearby golf course, before telling her to look up at the sky, where she saw the aurora borealis.

But the real turning point in Joan's becoming a scientist came when she was 14, and Richard was a graduate student at Princeton. Joan had long been fascinated by astronomy, but had actually been told by her mother that the female brain wasn't up to doing science.[13] Then, on her fourteenth birthday, Richard gave her a college level textbook on astronomy, and when she protested that it was too difficult for her, he told her to persevere. 'You start at the beginning and you read as far as you can, until you get lost. Then you start at the beginning again, and you keep working through until you can understand the whole book.'[14] Persevering in this way, she made steady progress. Eventually, she came to page 407, where there was a graph showing part of the spectrum of a star. The caption credited the astronomer who had obtained the information – Cecilia Payne-Gaposhkin – a woman! 'The secret was out: it was possible! From that day on, I was able to take my own interest in science seriously.'[15]

There was 'this excitement in the house, this great love of physics, so naturally I thought it sounded great', she remembers.[16] 'The feeling of excitement was in the house all the time, in my brother and my father. So I just grew up with it. Science became the thing to do.' But she was never any more in awe of Richard than other kid sisters were in awe of their big brothers. 'Your brother, he's your brother. You don't make any assumptions he's particularly brilliant.' It is only hindsight that made her realize that the family was actually unusual in its interest in science. 'Well, we were interested in relativity when I was a kid, so that then we had to be different than many other families.'

Two decades after Ritty had shown her the aurora, after she had finished her own PhD in solid state physics, Joan became interested in the aurora again. She was enjoying the work, and wanted to tell Richard about it. But the last thing she wanted was for her smart elder brother to solve the problem before she could have the pleasure of working it all out. So she went up to him and offered a deal, dividing up the Universe. If he would promise not to work on the aurora, she would leave everything else to him. Richard agreed.

In the 1980s, however, he visited Alaska, where he was shown around an observatory dedicated to the study of the aurora. Having learned about the work being done there, and expressing interest in the intriguing problems still to be solved, he was asked, well, why don't you work on some of these puzzles yourself? 'I would like to', Feynman replied, 'but I can't. I'd have to get my sister's permission.'

A little later, at a meeting of aurora experts, one of the Alaskan researchers came up to Joan, asking whether her brother had been joking. No, she said, the story was correct. On his return to California, Richard had asked her permission to work on the aurora, and she had turned him down. True to his word, given three decades earlier, he left the aurora to her.[17]

About the time Richard showed his little sister the aurora for the first time, he started in high school, in the autumn of 1931. By then, he was already established as an unusually clever child, both within school and outside. It was during the years in Cedarhurst that he really began to develop a conscious interest in science, and he was allowed to have a laboratory in the basement of the house, where he could experiment with chemicals. School in Cedarhurst, as far as science was concerned, was a complete waste of time. It was taught only in the eighth grade (the last grade in elementary school), and the only thing Feynman ever learned from it was that there are 39.37 inches in 1 metre. But in arithmetic, it was different. He was already 'known as some kind of a whiz-kid at arithmetic in elementary school', and at the age of 10 or 11 he was called out of his class and into another to explain his method of doing subtraction, which the teacher thought was particularly neat, to the younger children.[18]

In his last year at elementary school, though, Richard did make some of his first scientific contacts. He had a dentist who took the trouble to answer his questions about how teeth worked, and who he built up in his mind as 'a scientist'. He also tried to struggle through the few popular books in the public library about new developments in science (more of these developments in Chapter 2), and although the dentist was not really much of a scientist he realized that Richard had more than a passing interest in scientific matters. The dentist had another patient, William LeSur, who was an English teacher in Far Rockaway High School, but who helped out with the science teaching there; he told him about the boy's interest. The outcome was that LeSur invited Richard to visit the high school once a week, after classes had finished, and hang out in the lab while they cleaned up. Through this contact, Richard met the real chemistry teacher at the high school, and the head of science, Dr Edwin Barnes, who talked to him about science while he helped clean up the apparatus.

But if Richard learned little science from the teachers at Cedarhurst, it was during his time there that he learned about atoms from a new friend, Leonard Mautner, who explained what would happen if you kept on breaking up a substance into smaller and smaller pieces. To someone who has had any kind of scientific education, that may sound fairly trivial. But it was a landmark event in Feynman's life. Just over 30 years later, in his famous *Lectures*, he would say:

> If, in some cataclysm, all of scientific knowledge were to be destroyed, and only one sentence passed on to the next generations of creatures, what statement would contain the most information in the fewest words? I believe it is the *atomic hypothesis* (or the atomic *fact*, or whatever you wish to call it) that *all things are made of atoms – little particles that move around in perpetual motion, attracting each other when they are a little distance apart, but repelling upon being squeezed into one another.* In that one sentence, you will see, there is an *enormous* amount of information about the world, if just a little imagination and thinking are applied.[19]

Imagination and thinking were what the pre-teenage Richard Feynman (like the adult Richard Feynman) was superb at. In one of his favourite anecdotes (or parables, if you prefer), he told how while he was in Cedarhurst he learned how to repair radios. Radio sets were simple in those days, and he had started out, in Far Rockaway, by building his own crystal set, then moved on to fixing some problems for the family. Word spread, and friends and acquaintances used to call him in, rather than go to the expense of calling a regular radio repair man. The highlight of the story comes when a total stranger asks the kid to fix his radio, which makes an awful noise when it is switched on, but then settles down when it has warmed up. The kid paces up and down, trying to work out what is going on, while the owner of the radio gets more and more agitated, muttering about how stupid he has been to ask a little kid to do a man's job, and asking what Feynman is up to, to which the kid replies, 'I'm thinking.'

Eventually, having thought things through carefully, the kid realizes that the problem might be solved by reversing the order of two of the tubes (valves) in the radio. He swaps the tubes, switches it on, and it works perfectly. The owner of the set is enchanted, completely converted to the cause of the budding genius, and gets him more work, telling all his friends, 'He fixes radios by *thinking*!'[20]

Now, the point of the story is not that the older Feynman was on some ego trip, boasting about his childhood achievements. It is a story (which happens to be true) about the importance of imaginative thought, and how to solve problems in general. At another level, here is someone who was opposed to what Feynman was trying to do (or, at least, to the way he was trying to do it) who turned around completely to become almost embarrassingly enthusiastic once the technique had been shown to work. So when you know you are right, you should keep your courage in the face of opposition, carrying on the way you know is right. And it also tells us something a little more subtle about Feynman's character – he did not give up. Faced with a puzzle of any kind, from a neighbour's broken radio to the fundamental nature of quantum physics, he did not rest until he had solved it (unless, of course, he had promised his sister not to try).

In high school, the pattern continued. Older students would

come to him, for example, with tricky geometrical problems they had been assigned in the advanced mathematics class, and he would solve the puzzles – not because he was trying to ingratiate himself with the older boys, but because he couldn't resist the challenge. As it happens, the reputation he developed for being some kind of whiz at maths did help him socially. He was hopeless at ball games and what were generally regarded as 'manly' pursuits, shy with girls, and worried about being thought a 'sissy'. In *What Do You Care What Other People Think?* he describes being 'petrified' when passing a group of kids playing a ball game in case the ball rolled in his direction and he would be expected to pick it up and throw it back. The ball would always fly out of his hand in totally the wrong direction and everybody would laugh. The fact was, though, that he was simply too useful to the older boys for them to alienate him by making too much fun of these deficiencies.

Richard always tackled those geometry problems (and all other problems) his own way, using techniques that he had developed largely by himself, from first principles. Partly out of a desire to do it himself, partly through Melville's instruction that you shouldn't believe anything just because somebody else, no matter how eminent, told it to you, that was the way Feynman would work throughout his scientific life. With his friend Mautner, but largely on his own, he worked out most of the rules of Euclidean geometry for himself. 'I wanted to find the formula', he told Jagdish Mehra in 1988. 'I didn't care whether it had been worked out by the Greeks or even by the Babylonians; that didn't interest me at all. It was *my* problem, and I had fun out of it.'

He was also, as he put it, lucky enough to learn algebra his own way before coming into contact with it at school. His older cousin Robert could never get to grips with algebra, and had a tutor who came to coach him. Feynman was allowed to sit in on these sessions, and quickly learned that in algebra the problem was to find the value of the unknown variable, x, in an equation. While Robert struggled to do this by rote, using rules memorized at school, Feynman appreciated that it didn't matter *how* you got the answer, as long as it was the right one. Before he left elementary school, Richard had learned how to solve simultaneous equations – sets of two equations with two unknown quantities, such as

$$2x + y = 10$$
and
$$2y - x = 5$$

to find the values of both x and y (in this case, $x = 3$ and $y = 4$). Then, he made up for himself a problem with four equations and four unknowns.

Hardly surprisingly, by the time Richard came to algebra in high school he was bored to tears by what was on offer. He suffered in silence for a while, then told the teacher that he already knew what she was trying to teach the class. The head of the mathematics department gave him a problem to solve as a test; it was too difficult for him, but he made a good enough stab at it for them to see he really did know something about algebra. So he was put in a special class for the subject, really for students who had failed algebra once and were repeating it, with a teacher, Lillian Moore, flexible enough to cope with Richard's precocity. It was here that he met a new kind of puzzle. Miss Moore asked the class to solve the equation $2^x = 32$. Nobody could make head or tail of it. They didn't have a set of rules for solving that kind of problem. But Richard didn't need a set of rules; he saw straight away that the solution is $x = 5$, because 5 twos multiplied together is 32. This kind of thing was self-evident to Richard, and the fact that nobody else in the class felt the same way was one of the first indications he had that he really was different from the other students.

That difference came to the fore when Richard became the star of the school maths team, competing with other New York high schools in the 'Interscholastic Algebra League'. The algebra team would travel to different schools to compete with their maths whizzes. There were five members of each team, and they would be given problems that required what would nowadays be called lateral thinking to solve, with a strictly limited time in which to solve them – typically 45 seconds. Each member of the team worked independently, and could write anything he wanted on the paper in front of him. All that mattered was that before the time was up each competitor had to draw a circle around the one number on the paper that was his answer to the problem. The problems were deliberately chosen so that although they could, of course, be solved 'by the rulebook', it would be just about impossible to do so in the time available; but they were easy once you saw the short cut (or invented your own short cut). Feynman always won these competitions, writing down his number and ostentatiously drawing a circle around it, often on an otherwise blank piece of paper, usually before the other competitors had really got to grips with it at all. The practice served him well in later life, when he retained the ability to solve algebraic problems quickly and neatly, without ploughing through the textbook methods.

So Richard learned a little maths in high school, although he always claimed that he didn't learn any science at all there, because he was always ahead of what was being taught in class. The kind of biology, physics and chemistry taught in Far Rockaway High School in the 1930s was already familiar to him from the *Encyclopaedia Britannica*, his own tinkering (for example with electricity), and informal conversations with his teachers and others. Even the maths he learned while at high school was largely self-taught – the big new thing for him in those years was calculus, which he learned from two books, *Calculus Made Easy*, by S. P. Thompson (St Martin's Press, New York, 1910) and *Calculus for the Practical Man* by J. E. Thompson (Van Nostrand, New York, 1931), one of a series of 'practical man' guides to mathematics that Richard devoured around the time he left elementary school and went to high school.

But two mathematical experiences that Richard had while in high school did stick with him for the rest of his life. One gave him an insight into what it was like for ordinary students; the other shaped his entire subsequent career.

The glimpse of mathematical mortality came when Richard was introduced to solid geometry, the study of shapes in three dimensions, in high school. He was completely thrown, and couldn't understand what the teacher was getting at at all, although he could use the rules the teacher gave in order to carry through calculations properly. For once, he was in the same position as students who used the rules of algebra to solve equations without understanding what was going on. Then, the penny dropped. After a couple of weeks, he realized that the mess of lines being drawn on the blackboard was indeed meant to represent three-dimensional objects, not some crazy pattern in two dimensions. Everything came into focus, and he never had any trouble with the subject again. As far as science was concerned, 'it was my only experience of how it must feel to the ordinary human being', he later said.[21]

In 1933, the Feynman family visited the World's Fair in Chicago; a year later, Richard began his final year in high school, and made the mathematical encounter that was to shape his career.

He owed the encounter to the Depression. That year, a new physics teacher, Abram Bader, joined the school. He had been working for a PhD at Columbia University, under the Austrian-born physicist I. I. Rabi, whose work on the magnetic properties of fundamental particles would bring him the Nobel Prize in 1944. But Bader ran out of money, and had to drop out of research to become a teacher. He quickly appreciated Feynman's unusual abilities, lending him a book on advanced calculus, and often talking to him, out of class, about scientific matters. Once he explained something called the Principle of Least Action. They discussed the topic only once, but the whole scene stuck in Feynman's mind for the rest of his life. He was so excited by the idea that he remembered everything about the occasion – exactly where the blackboard was, where he was standing, where Mr Bader was standing, and the room they were in. 'He just explained, he didn't prove anything. There was nothing complicated; he just explained that such a principle exists. I reacted to it then and there, that this was a miraculous and marvelous thing to be able to express the laws in such an unusual fashion.'[22]

The 'miraculous and marvelous thing' can be understood in terms of the flight of a ball tossed from the ground through an upper-storey window. In this context, the term 'action' has a precise meaning. At any point in its flight, you can calculate the difference between the kinetic energy of the ball (the energy of the ball's motion, related to its speed) and its potential energy (the gravitational energy the ball possesses because of its height above the ground). The action is the sum of all these differences, all along the path of the ball through the air (action can be calculated in a similar way for charged particles moving in electric or magnetic fields, including electrons moving in atoms). There are many different curves the ball could follow to get through the window, ranging from low, flat trajectories to highly curved flight paths in which it goes far above the window before dropping through it. Each curve is a parabola, one of the family of trajectories possible for a ball moving under the influence of the Earth's gravity. All this Feynman knew already. But Bader reminded him that if you know how long the flight of the ball takes, from the moment it leaves the thrower's hand to the moment it reaches the window, that rules out all but one of the trajectories, specifying a unique path for the ball. And then he told him about the Principle of Least Action.

One of the most important principles in physics is the conservation of energy – the total amount of energy associated with the ball (in this example) stays the same. Some of this energy is in the form of gravitational potential energy, which depends on its height above the surface of the Earth (strictly speaking, on its distance from the centre of the Earth). When the ball rises, it gains gravitational potential energy; when it falls, it loses some of this energy. The only other relevant form of energy possessed by the ball is its energy of motion, or kinetic energy. Higher speeds correspond to greater kinetic energy. At the moment the ball leaves the thrower's hand, it has a lot of kinetic energy because it is moving fast. As it rises, some of this kinetic energy is lost, traded for gravitational potential energy, and it slows down. At the top of its trajectory, it has minimum kinetic energy and maximum potential energy, then as it falls down the other side of the curve it gains kinetic energy and loses potential energy. But the *total*, the sum of (kinetic + potential) energy is always the same.

All this Feynman knew. But what he didn't know was that given the time taken for the journey, the trajectory followed by the ball is always the one for which the *difference*, kinetic energy *minus* potential energy, added up all along the trajectory, is the *least.* This is the Principle of Least Action, a property involving the whole path.

Looking at the curved line on a blackboard representing the flight of the ball, you might think, for example, that you could make it take the same time for the journey by throwing it slightly more slowly, in a flatter arc, more nearly a straight line; or by throwing it faster along a longer trajectory, looping higher above the ground. But nature doesn't work that way. There is only one possible path between two points for a given amount of time taken for the flight. Nature 'chooses' the path with the least action – and this applies not just to the flight of a ball, but to any kind of trajectory, at any scale. Mr Bader didn't work out the numbers involved, or ask Feynman to work them out. He just told him about the principle, a deep truth which impressed the high school student in his final year before going on to college.

It's worth a slight detour to give another example of the principle at work, this time in the guise of the Principle of Least Time, because it is so important both to science and to Feynman's career. This version of the story involves light. It happens that light travels slightly faster through air than it does through glass.* Either in air or glass, light travels in straight lines – an example of the Principle of Least Time, because, since a straight line is the shortest distance between two points, that is the quickest way to get from A to B. But what if the journey from A to B starts out in air, and ends up inside a glass block? If the light still travelled in a single straight line, it would spend a relatively small amount of time moving swiftly through air, then a relatively long time moving slowly through glass. It turns out (see Figure 1) that there is a unique path which enables the light to take the least time on its journey, which involves travelling in a certain straight line up to the edge of the glass, then turning and travelling in a different straight line to its destination. The light seems to 'know' where it is going, apply the Principle of Least Action, and 'choose' the optimum path for its journey.

The connection between mathematics and physics highlighted by the Principle of Least Action reinforced a growing fascination that Richard had had with this area of science right through high school. While working with radio receivers, building his own circuits and working out how to tune them, he had come across equations describing the behaviour of these practical objects that involved the Greek pi, the ratio of the circumference of a circle to its diameter. Although there were circular (or cylindrical) coils in these circuits, it is also possible to work with square coils, and pi came into the equations whatever the shape of the coils. There was some deep link between physics and mathematics, which Feynman did not understand, but which intrigued him. Although still

*The famous 'ultimate speed limit' from relativity theory is the speed of light *in a vacuum*, which is greater still.

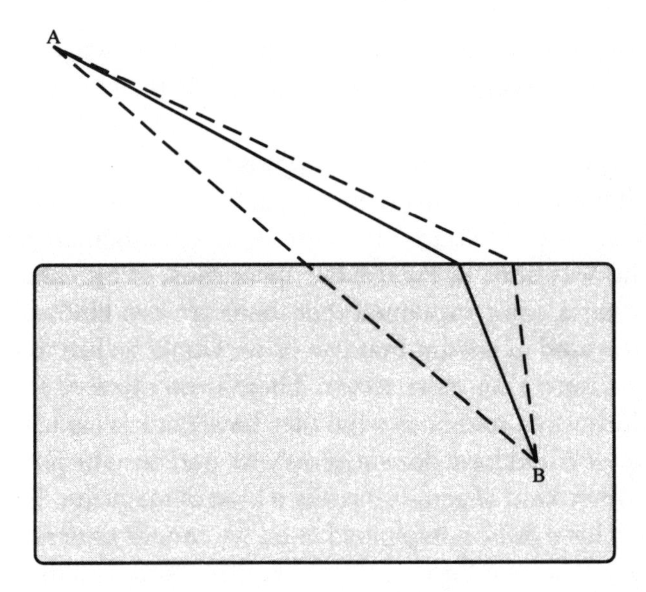

Figure 1. Light travels faster through air than through glass. So the quickest journey from A to B that is partly through air and partly through glass is not the (dotted) straight line from A to B, but there is a unique 'path of least time' made up of two straight lines. This is a special case of the Principle of Least Action at work. The dotted lines to the right show an example of a path that takes longer than the path of least time (solid lines).

known as a whiz at maths, his fascination was really with physics.

We have emphasized the role of science in young Richard's life because it was, indeed, the main thing in his life. He went through the educational system in what seemed, superficially, a conventional way, but actually learned his science for himself, outside the system (including teaching himself about relativity theory from books while still in high school). He found school boring, but sailed through examinations with ease, appearing, in that respect, to have been a model student.

How clever did he have to be, to do all that? Joan Feynman once sneaked a look at the results of the standard IQ tests that both she and her brother had taken in high school.[23] Her score was 124, his was 123, so she could always claim to be smarter than he was. It is notoriously true that IQ tests are only any good at measuring the ability of people to do IQ tests, but much later in life Feynman took great delight in being able to quote his IQ score when invited to join the organization Mensa, which is exactly the kind of 'club' for the self-important that he despised. Unfortunately, he replied, he could not join Mensa because his IQ was not high enough for them.

But that didn't stop his being a genius, because some kinds of genius cannot be measured in IQ tests. The mathematician Mark Kac, who was born in Poland but spent most of his career in the United States, once explained that there are two kinds of genius. One is the kind of person that you or we would be just as good as, if only we were a lot more clever. There is no mystery about how their minds work, and once what they have done is explained to us we think we could have done it, if only we had been bright enough. But the other kind of genius is really a kind of magician. Even after what they have done is explained to us, we cannot understand how they did it. 'Richard Feynman', said Kac in 1985, 'is a magician of the highest caliber.'[24]

But in spite of being clearly different from his peers in this way, even as a child, and in spite of his fears of being thought a sissy, Richard wasn't what would now be called a 'nerd'. He had a handful of close friends, some interested in science, others on the humanities side; and his feet were kept firmly on the ground in those Depression days by the need to work at odd jobs to earn spending money.

His father earned about $5000 a year in the early 1930s, which Richard knew because Melville would sometimes send him to the bank with a cheque for a week's salary, about $100. It was Melville's way of teaching Ritty the value of money, and it worked. Richard knew how much money the family had, and knew that they lived in reasonable comfort. He remembered thinking 'that everything was all right, we lived fine, and my ambition was to earn that much money . . . I knew I wanted about $5000 a year, that's all I needed.'[25] He worked part time for a printer while in high school, and in the summer after graduation at a hotel run by his aunt, giving rise to some of the stories recounted in *Surely You're Joking, Mr. Feynman!* In many ways, this was a typical lifestyle for a bright Jewish kid in New York in the 1930s – it bears striking resemblances, for example, to the story Isaac Asimov tells in his autobiographical *I Asimov* (Doubleday).

As a teenager, Feynman later said, he was interested in only two things, maths and girls (that's one more thing than most teenage boys are interested in). He learned to dance, which was very useful for one of his interests, and he also quickly learned about the difference between social niceties and the truth.

Richard was always uncomfortable with the phoney way many other people used language. He regarded English, as taught in high school, as 'a kind of baloney', had a lifetime disdain of philosophy, and dismissed religion, which seemed to him to be based purely on wishful thinking. He talked straight, meant what he said, and was genuinely confused if that seemed to upset other people. So when, at the end of his first date, the girl he had taken out said, 'Thank you for a very lovely evening', he thought she meant it. When the next girl he took out ended the evening with exactly the same words, he began to wonder. So the third time he took a girl out on a date, when the time came to say goodnight he got in first, saying the stock phrase before she could, and leaving her tongue-tied, unable to think what to say in response, because she had been just about to say the same thing.[26] This was one of Feynman's first encounters with this kind of empty formality, but he seldom bothered with such niceties himself.

It wasn't too long before he got to know a girl who would end up caring even less about social niceties than he did, and making him blissfully happy, as his first wife, in the process. When Richard first met Arline Greenbaum, when he was about 13, she was one of his wider circle of acquaintances, not a close friend. As they grew up together, he got to dance with her on occasion, but she soon had a regular boyfriend and to a large extent he admired her from a distance (Joan Feynman recalls Richard first mentioning this 'wonderful girl' to her when he was about 15 and Joan was six). Arline was the most popular girl in the group, and everybody liked her. As Feynman recounted in *What Do You Care What Other People Think?*, she once made his day simply by coming over to him at a party and sitting on the arm of his chair to talk to him. 'Oh boy!', he thought, 'somebody I like has paid attention to me!' (his comments at home the next day may have been the occasion Joan remembers). He even joined an art group, something he had no ability at whatsoever at that time, simply because Arline was a member.

Eventually, Arline's steady relationship with her boyfriend

ended, and during his final year in high school Richard got to know her better, although she was still dating other boys at that time. But Harold Gast, one of Richard's contemporaries who also dated Arline, says that by then it was obvious to everyone in the group 'that they were really very fond of each other and nobody was going to interfere'.[27] Still rather shy and insecure socially, however, Richard imagined that Gast was a serious competitor, and was relieved when Arline chose to sit with Melville and Lucille at his graduation ceremony, a public acknowledgement of her interest in him.

The graduation was, of course, a triumph for Feynman, who took top honours in just about everything, ironically including English. The reason for that particular triumph, also recounted in *What Do You Care*, was that, knowing his limitations, in the examination he had written an uncontentious essay about technology and aviation, designed to appeal to his teachers by 'slinging the bull' – saying simple things in an impressive way, using long words and technical terms. His friends with greater literary talent (including Gast) had been confident enough to spread their wings and take up more controversial themes, with which the examiners could take issue (another example, to Feynman, of the 'baloney' involved in English). So they 'only' scored 88 per cent, while Richard scored 91 per cent (in one of his worst subjects).

In those days (Feynman graduated in the summer of 1935) many bright kids had to forgo a college education for financial reasons, but Melville and Lucille were determined to give Richard the best education they could. Even with his academic track record and his parents' backing, though, getting into college wasn't all plain sailing. He applied to Columbia University and the Massachusetts Institute of Technology (MIT). Columbia required an examination, and charged would-be students $15 for the privilege of taking it (at a time when the Feynman family income, remember, was about $100 per week); Richard took the exam, and presumably passed, but was denied a place at Columbia because they had already filled up their quota of Jewish students for that year. Feynman wasn't bothered by the quota system, incredible though it seems to modern eyes; that was just the way things worked in the 1930s. But he would probably have appreciated it if the university had rejected him out of hand, without taking his $15 first.

That left MIT. Apart from the academic requirements, they insisted upon a recommendation from an MIT graduate for all prospective freshmen. This did rankle, but it was a hoop that had to be jumped through, and Melville did the jumping, persuading an acquaintance whom he knew had gone to MIT to provide the recommendation. But the acquaintance really knew nothing about Richard, who later described the system,[28] as 'evil, wrong, and dishonest', a falseness that was the only thing he disliked about applying to MIT. The unpleasant taste was eased somewhat when the college offered Richard a small scholarship – he had applied for a full scholarship, which he failed to get, but received the small award of about $100 per year, which would be a help.

In the summer of 1935, before he left for MIT, Feynman worked in his aunt's hotel (putting money aside ready for college) and spent a lot of time getting to know Arline better. It was at MIT that he would formally make the transition from being a mathematician to being a physicist, and he was lucky enough to arrive on the scene at a time when the physics textbooks had been completely rewritten by the development, in the 1920s, of quantum theory. The younger Feynman had read about some of this new work already, for pleasure; soon, it would become his vocation. In order to appreciate where Feynman was coming from when he began to make his own original contributions to science, it is time to take stock of the state physics was in just before Feynman came on the scene, in the aftermath of the quantum revolution and the slightly older revolution initiated by Albert Einstein with his two theories of relativity. Twentieth-century science was a very different world from the one in which physicists had operated for the previous 200 years, from the time of Isaac Newton (at the end of the seventeenth century), to the time of Max Planck (at the end of the nineteenth century).

Notes

1. Richard Feynman, interview with Jagdish Mehra, quoted in Mehra's book *The Beat of a Different Drum* (hereafter referred to as Mehra; details in bibliography).

2. Feynman often recounted this anecdote. See, for example, *What Do You Care What Other People Think?*, by Richard Feynman & Ralph Leighton (hereafter referred to as *What Do You Care*; details in Bibliography). The widely recounted dinosaur, bird and wagon anecdotes can be found in the same source.
3. *What Do You Care.*
4. Leighton, interview with JG, April 1995.
5. Quoted in *No Ordinary Genius*, edited by Christopher Sykes (see Bibliography).
6. *What Do You Care.*
7. The story of Johanna and Henry Phillips is told by Joan Feynman in *No Ordinary Genius.*
8. *What Do You Care.*
9. See Joan Feynman's contribution to the Feynman memoir *Most of the Good Stuff*, edited by Laurie Brown & John Rigden (see Bibliography).
10. Joan Feynman's comments taken from correspondence with JG, January/February 1996.
11. Mehra.
12. *Most of the Good Stuff.*
13. There is no evidence that Lucille believed this. But she must have been aware of the extremely limited career opportunities for women in science at the time, and was probably trying to steer Joan away from the likelihood of a major disappointment.
14. Interview with JG, April 1995; see also *No Ordinary Genius.*
15. *Most of the Good Stuff.*
16. Interview with JG, April 1995.
17. *No Ordinary Genius*; see also note 10.
18. Mehra.
19. See also *Six Easy Pieces* (see Bibliography).
20. See, for example, *Surely You're Joking, Mr. Feynman!*, by Richard Feynman & Ralph Leighton (see Bibliography; hereafter referred to as *Surely You're Joking*).
21. Mehra.
22. Mehra.
23. *No Ordinary Genius.*
24. Quoted by Hans Bethe, whom Kac described as an ordinary genius, in *No Ordinary Genius.*
25. Mehra.
26. *What Do You Care.*
27. Mehra.
28. Mehra.

— 2 —

Physics before Feynman

The two revolutions that transformed physics in the twentieth century, relativity theory and quantum mechanics, both developed from new understandings of the nature of light, and both had their roots in the nineteenth century. When Albert Einstein developed his Special Theory of Relativity early in the twentieth century[1] (it was published in 1905), the foundation stone on which he built was a discovery that had been made four decades earlier, in the 1860s, by the Scottish physicist James Clerk Maxwell.

Maxwell, who was born in 1831 and died in 1879 (the year Einstein was born), was one of the great physicists of his day, who made many contributions to science. But he is best remembered for his work on electricity and magnetism, which led him to the discovery that light can be described as an electromagnetic wave travelling through space at a certain speed. He developed a set of four equations, now known as Maxwell's equations, which can provide the answer to any question you want to ask about the 'classical' (that is, pre-quantum theory) behaviour of electricity and magnetism. Maxwell's equations will tell you the force that operates between two electrical charges of a certain strength a certain distance apart; they will tell you how strong an electric current is generated in a nearby wire by a magnet moving past at a certain speed; and so on. Every problem involving electricity and magnetism, above the quantum level, can be solved by using Maxwell's equations, which represented the greatest unifying discovery in science since Isaac Newton discovered the Universal Law of Gravitation.

One solution of Maxwell's equations, a natural component of the unified whole, describes electromagnetic waves moving through space. The speed with which the waves move, usually denoted by the letter c, is a constant which emerges naturally from the equations, as a fundamental property of nature. It is *not* put in by hand. It was when Maxwell found that the value of c which automatically comes out of his theory is exactly the same as the speed of light measured in a vacuum (which was already quite well determined by the 1860s) that he realized that his equations also described the behaviour of light. In 1864, he wrote:

> The velocity is so nearly that of light that it seems we have strong reason to conclude that light itself . . . is an electromagnetic disturbance in the form of waves propagated through the electromagnetic field according to electromagnetic laws.[2]

That word 'field' is one to watch out for. It is related to the idea of lines of force, which helps us to visualize, for example, what happens when two magnets are brought together. In this case, the lines of force are thought of as something like stretched elastic bands, which start out from the magnetic 'north pole' on a bar magnet and end up on the magnetic 'south pole'. When a north pole and a south pole are brought together, the lines of force reach out across the gap and pull the two poles together; but when two north poles are pushed together, the lines of force are forced out of the gap, creating a resistance and holding the two north poles apart (see Figure 2). The region around the magnet where it exerts this influence is the region of its 'magnetic field'. In a similar way, physicists think of massive objects, like the Sun and the Earth, as being surrounded by a 'gravitational field', filled with lines of force that tug on any object in that field. Of course, lighter objects, such as our desk, or your pen, also have their own gravitational fields, but these are so weak that they can only be detected using very sensitive equipment.

Field theory is an extremely successful way of describing the interactions between things like magnets, electrical charges and gravitating bodies. But don't run away with the idea that it is the *only* way to describe these interactions. Without wishing to get too far ahead of our story, it's worth warning you that one of the things that most intrigued Richard Feynman in later life was the way in which several different descriptions of the way things work can turn out to be equally effective in the right hands. Maxwell himself actually worked towards his field theory through an intermediate image

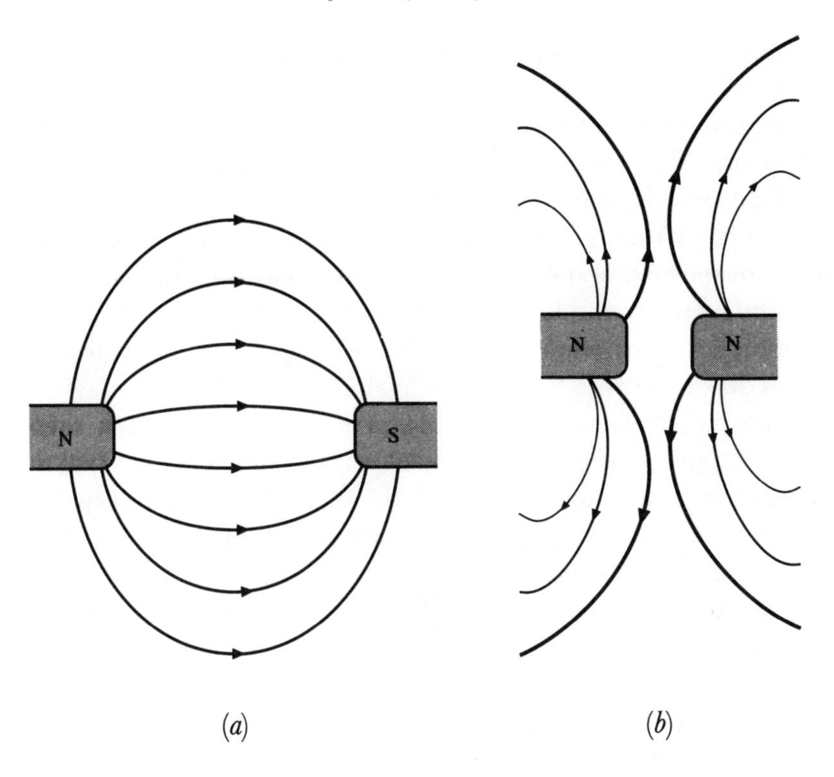

Figure 2. The concept of a field is related to the idea of 'lines of force'. (*a*) A north magnetic pole and a south magnetic pole attract each other as if they were being pulled together by stretched elastic bands. (*b*) Two north magnetic poles repel each other as if they were separated by a block of stiff, compressed rubber. The magnetic field is stronger where the lines of force are closer together.

which involved the forces of electricity and magnetism being conveyed by whirlpool-like vortices spinning in a fluid which filled all the space between material objects. The way the vortices interacted was like the cogs and wheels of some great piece of clockwork, and this early version of the theory looks totally bizarre to modern eyes – but it worked. The lesson to be drawn is that in some deep sense the truth about how the world works resides in the equations – in this case, Maxwell's equations – and not in the physical images that we conjure up to help our limited imaginations to visualize what is going on.

That was a point that was well appreciated by the young Einstein.

One of the strangest things about the constant c that appeared in Maxwell's equation was that it was just that – a *constant.* It represented the speed of light (and all other electromagnetic radiation, including radio waves), but it took no account of how fast the object producing the light was moving, or how fast the person measuring the speed of the light was moving. This didn't match common sense, or the laws of motion based upon Newton's work in the seventeenth century and held sacrosanct ever since.

In the everyday world, if you ride in an open car that is travelling at 50 kilometres an hour (km/h) along a straight road, and you throw a ball straight out ahead of you at a speed of 5 km/h, then (if you could ignore wind resistance) you would expect the ball to be moving at 55 km/h relative to the road. But what Maxwell's equations seemed to say was that if you rode in the same car and shone its headlights out in front of you, the speed of the light from the car would not only be c relative to the car (as you would expect) but also c (not c + 50 km/h) relative to the road! Even if you were in a spaceship travelling at half the speed of light, and you met a spaceship travelling the opposite way at half the speed of light, the light from the headlights on the other spaceship would be travelling at the same speed of light, c, relative to your measuring instruments *and* relative to the measuring instruments in the other spaceship.

It was clear by the end of the nineteenth century that there must be something wrong either with Maxwell's equations or with common sense (and Newton's equations). It was Einstein's genius to take Maxwell's equations at face value, and work out all the implications in his Special Theory of Relativity. Einstein's theory explains how it can be that the speed of light (in a vacuum; it travels slightly more slowly in more dense media) is always measured to be the same no matter how the measuring instruments are moving relative to the light source. The implications include the fact that the faster an object moves, the more massive it gets; the fact that nothing can be accelerated from 'ordinary' speeds to travel faster than light (so that even if you are in a spaceship travelling at two-thirds of c relative to Earth, and you encounter a spaceship travelling in the opposite direction at two-thirds of c relative to Earth, the velocity of the other spaceship relative to yours is still less than c); and the famous relationship between mass and energy, $E = mc^2$.

All of these predictions, it cannot be overemphasized, have been tested many times to great precision. The Special Theory of Relativity passes every test, and has been proven to be a good description of the way the world works.[3] But you only need to use the Special Theory to understand what is going on if you are dealing with things moving at very high speeds, a sizeable fraction of the speed of light. The difference between the predictions of the Special Theory and common sense are of no significance at all for speeds that are small compared with the speed of light, which is itself a huge 300,000 kilometres per second. Unfortunately for the physicists, though, there are things which move at these so-called 'relativistic' speeds that have to be taken account of in their attempts to describe the way the everyday world works. In particular, electrons whizzing around inside atoms have to be described taking proper account of the Special Theory of Relativity.*

By the time Feynman went to MIT, the structure of the atom, and the way it operated in accordance with both quantum mechanics and special relativity, were pretty well understood, except for some annoying details. The electron had been identified in the 1890s by the British physicist J. J. Thomson, the role of the proton was appreciated by the beginning of the 1920s, and the neutron was identified in 1932. This combination of particles was all that was needed to explain the structure of atoms. Each atom contains a nucleus that is a ball of positively charged protons and electrically neutral neutrons, held together (in spite of the tendency of the positive charge on the protons to make them repel one another) by a very short range force of attraction, called the strong nuclear force. Outside the nucleus, each atom 'owns' a cloud of electrons, with one negatively charged electron for each proton in the nucleus, held in place by the mutual attraction between the negative charge on the electrons and the overall positive charge on the nucleus. In addition, during the early 1930s physicists began to suspect the existence of another type of particle, dubbed the

*Einstein's second great theory, the General Theory of Relativity, is a field theory of gravity, and is quite different (in spite of the similarity of names) from the Special Theory of Relativity. It comes into our story later, but it had little bearing on the mainstream of physics research in the 1930s and 1940s.

neutrino, which had never been detected directly but was required to balance the energy budget whenever a neutron transformed itself into a proton by spitting out an electron (a process known as beta decay). Beta decay involves a fourth kind of force (after gravity, electromagnetism and the strong force), dubbed the weak force, or weak interaction.

Together with light, that's all you need to explain the workings of the everyday world. But to anyone brought up on classical ideas (the kind of physics you get taught in school), there's an obvious puzzle about this picture of the atom. Why don't all the negatively charged electrons in the outer part of the atom get pulled into the nucleus by the attraction of all the positively charged protons? The world would be a far different place if they did, because the nucleus is typically about 100,000 times smaller than the electron cloud that surrounds it. The nucleus contains almost all of the mass of an atom (protons and neutrons have roughly the same mass, each about 2000 times the mass of an electron), but the electrons are responsible for the atom's relatively large size, and for the 'face' it shows to the world (that is, to other atoms). The reason they don't fall into the nucleus is explained by the second revolution in twentieth-century physics, the quantum revolution. Like the relativity revolution, this was also triggered by studies of the behaviour of light.

At the end of the nineteenth century, the world seemed to be made up of two components. There were particles, like the newly discovered electrons, and there were waves, like the electromag-

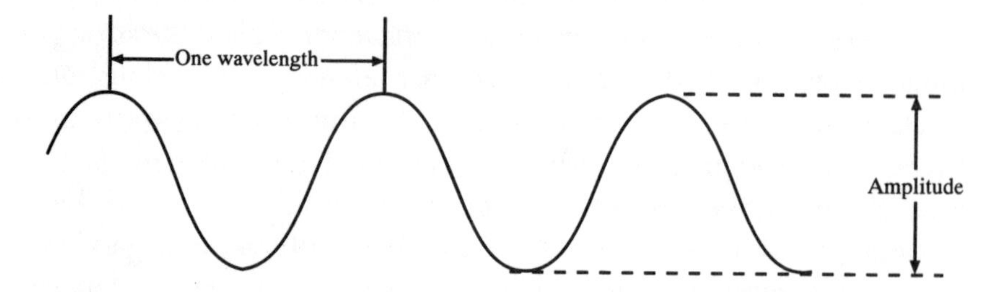

Figure 3. A wave. Two waves are in phase if they move in step so that the peaks reinforce one another. They are out of phase if the peaks of one wave exactly coincide with the troughs of the other wave, so that they cancel each other out. In-between states, with partial cancellation, are also possible.

netic waves described by Maxwell's equations. You can make waves in a bowl of water by jiggling your fingers about in the water, and you can make electromagnetic waves by jiggling an electrically charged particle to and fro. So it was pretty clear, even then, that light was produced by electrons jiggling about in some way inside atoms. Unfortunately, though, the best nineteenth-century theories predicted that this jiggling would produce a completely different spectrum of light from what we actually see.

What the theorists had to do was to explain the way light would be emitted from an idealized source called a 'black body'. This seemingly bizarre choice of name (if it is black, how can it radiate any light at all?) results from the fact that when such an object is cold, it absorbs all the light that falls on it, without reflecting any away. It treats all colours (each colour corresponds to a particular wavelength of light) the same. But if it is gradually heated up, it will first begin to radiate invisible infrared radiation, then it will start to glow red, then orange, yellow and blue at successively higher temperatures, until eventually it is white hot. You can tell the temperature of a black body precisely by measuring the wavelength of the light it is emitting. This light forms a continuous spectrum (the 'black body curve'), with most energy radiated in a peak at the characteristic wavelength for that temperature (corresponding to red light, or blue light, or whatever) but some energy coming out in the form of electromagnetic waves with shorter wavelengths than this peak intensity, and some with longer wavelengths. The shape of the black body curve is like the outline of a smooth hill, and the peak itself shifts from longer wavelengths to shorter ones as the black body gets hotter.

But according to nineteenth-century physics, none of this should happen. If you try to treat the behaviour of electromagnetic waves in exactly the same way that you would treat vibrations of a guitar string, it turns out that it ought to be easier for an electromagnetic oscillator to radiate energy at shorter wavelengths, regardless of its temperature – so easy, in fact, that all of the energy put into a black body as heat should come pouring out as very short wavelength radiation, beyond the blue part of the spectrum, in the ultraviolet. This was known as the 'ultraviolet catastrophe', because the prediction certainly did not match up with the real world, where such things as red hot pokers (which behave in some ways very much like black bodies) were well known to the Victorians.

The puzzle was resolved – up to a point – by the German physicist Max Planck, in the last decade of the nineteenth century. Planck, who lived from 1858 to 1947, spent years puzzling over the nature of black body radiation, and eventually (in 1900), as a result of a mixture of hard work, insight and luck, came up with a mathematical description of what was going on. Crucially, he was only able to find the right equation because he knew the answer he was looking for – the black body curve. If he had simply been trying to predict the nature of light radiated from a hot black body, he would never have produced the key new idea that did actually appear in his calculations.

Planck's new idea, or trick, was to assume that the electric oscillators inside atoms cannot emit any amount of radiation they like, but only lumps of a certain size, called quanta. In the same way, they would only be able to absorb individual quanta, not in-between amounts of energy. And in order to make Planck's formula match the black body curve, the amount of energy in each quantum had to be determined by a new rule, relating the energy of the quantum involved to the frequency (f) of the radiation. Frequency is just one over the wavelength, and Planck found that for electromagnetic radiation such as light

$$E = hf$$

where h is a new constant, now known as Planck's constant.

For very short wavelengths, f is very big, so the energy in each quantum is very big. For very long wavelengths, f is very small and the energy in each quantum is small. This explains the shape of the black body curve, and avoids the ultraviolet catastrophe. The total amount of energy being radiated at each part of the black body spectrum is made up of the contributions of all the quanta being radiated with the frequency (and wavelength) corresponding to that part of the spectrum. At long wavelengths, it is easy for atoms to radiate very many quanta, but each quantum has only a little energy, so only a little energy is radiated overall. At short wavelengths, each quantum radiated carries a lot of energy, but very few atoms are able to generate such high-energy quanta, so, again, only a little energy is radiated overall. But in the middle of the spectrum, where medium-sized quanta are radiated, there are many atoms which each contain enough energy to make these quanta, so the numbers add up to produce a lot of energy – the hill in the black body curve. And, naturally, the wavelength at which the peak energy is radiated shifts to shorter wavelengths as the black body gets hotter and more atoms are able to produce higher-energy (shorter wavelength) quanta.

Although physicists were pleased to have a black body formula that worked, at first this was regarded as no more than a mathematical trick, and there was no suggestion (least of all from Planck himself) that light could only exist in little lumps, the quanta. It took the genius of Albert Einstein to suggest, initially in 1905, that the quanta might be real entities, and that light could just as well be described as a stream of tiny particles as by a wave equation. Although Einstein's interpretation of the quantum idea neatly solved an outstanding puzzle in physics (the way in which light shining on a metal surface releases electrons in the photoelectric effect), initially it met with a hostile reaction. One American researcher, Robert Millikan, was so annoyed by it that he spent 10 years trying to prove Einstein was wrong, but succeeded only in convincing himself (and everybody else) that Einstein was right.

After Millikan's definitive experimental results were published (in 1916) it was only a matter of time before first Planck (in 1919) and then Einstein (in 1922, although it was actually the prize from 1921 held over for a year) received the Nobel Prize for these contributions. But the 'particles of light' were only given their modern name, photons, in 1926, by the American physicist Gilbert Lewis. By then, the Indian physicist Satyendra Bose had shown that the equation describing the black body curve (Planck's equation) could actually be derived entirely by treating light as a 'gas' made up of these fundamental particles, without using the idea of electromagnetic waves at all.

So, by the mid-1920s, there were two equally well-founded, accurate and useful ways of explaining the behaviour of light – either in terms of waves, or in terms of particles. But this was only half the story. We still haven't explained why electrons don't fall into the nucleus of an atom.

The first step, producing a picture of the structure of the atom that is still the one often taught in schools, was taken by the Dane Niels Bohr, in the second decade of the twentieth century. Bohr had been born in 1885 and lived until 1962. He completed his PhD studies in 1911 and a year later began a period of work in Manchester, where he stayed until 1916, working in the group headed by the New Zealand-born physicist Ernest Rutherford.

Bohr's model of the atom was like a miniature Solar System. The nucleus was in the middle and the electrons circled around the nucleus in orbits rather like the orbits of the planets around the Sun. According to classical theory, electrons moving in orbits like this would steadily radiate electromagnetic radiation away, losing energy and very quickly spiralling into the nucleus. But Bohr guessed that they could not do this because, extending Planck's idea, they were only 'allowed' to radiate energy in distinct lumps, the quanta. So an electron could not spiral steadily inwards; instead it would have to jump from one stable orbit to another as it lost energy and moved inward – rather as if the planet Mars were suddenly to jump into the orbit now occupied by the Earth. But, Bohr said, the electrons could not all pile up in the innermost orbit (like all the planets in the Solar System suddenly jumping into the orbit of Mercury) because there was a limit on the number of electrons allowed in each orbit. If an inner orbit was full up, then any additional electrons belonging to that atom had to sit further out from the nucleus.

The picture Bohr painted was based on a bizarre combination of classical ideas (orbits), the new quantum ideas, guesswork and new rules invoked to explain why all the electrons were not in the same orbit. But it had one great thing going for it – it explained the way in which bright and dark lines are produced in spectra.

Most hot objects do not radiate light purely in the smooth, hill-shaped spectrum of a black body. If light from the Sun, say, is spread out using a prism to make a rainbow pattern, the spectrum is seen to be marked by sharp lines, some dark and some bright, at particular wavelengths (corresponding to particular colours). These individual lines are associated with particular kinds of atoms – for example, when sodium atoms are heated or energized electrically they produce two bright, yellow–orange lines in the spectrum, familiar today from the colour of certain street lamps. Bohr explained such lines as the result of electrons jumping from one orbit (one energy level) to another within the atoms. You can think of this as like jumping from one step to another on a staircase. A bright line is where identical electrons in many identical atoms (like the sodium atoms in street lights) have all jumped inward by the appropriate step, each releasing the same amount of electromagnetic energy in the form of many quanta of light each with the same frequency given by Planck's formula $E = hf$. A dark line is where background energy has been absorbed by electrons making the appropriate jump up in energy, outward from one stable orbit into a more distant stable orbit ('up a step' on the staircase).

But why should only some orbits be stable, and others not? It was this puzzle that led the French physicist Louis de Broglie to make the next breakthrough in quantum theory, in the 1920s.

De Broglie, who was born in 1892, only began serious scientific work after his military service during the First World War and completed his PhD in 1924, at the relatively ripe old age of 32 (he lived to an even riper old age, until 1982). De Broglie suggested that the way in which electrons could only occupy certain orbits around a nucleus was reminiscent of the way waves behaved, rather than particles. If you pluck an open violin string, for example, you can make waves on it in which there are exactly 1, or 2, or 3, or any whole number of wavelengths, corresponding to different notes (harmonics) 'fitting in' to the length of the string, by lightly touching the string at various points that are simple fractions ($\frac{1}{2}$, $\frac{1}{3}$, $\frac{1}{4}$ and so on) of the length. But you can't make a note corresponding to a wave with, say, 4.7 wavelengths filling the open string. In order to play that note you have to change the length of the string by pressing it hard with your finger against the neck of the violin. If electrons were really waves, said De Broglie, then each orbit in an atom might correspond to patterns in which a whole number of electron waves fitted around the orbit, making a so-called standing wave. The transition from one step on the energy level staircase to another would then correspond more to the transition from one harmonic to another than to a particle jumping from one orbit to another.

De Broglie's suggestion was so revolutionary that his thesis supervisor, Paul Langevin, didn't trust himself to decide on its merits, and sent a copy to Einstein, who responded that he thought the work was reliable. De Broglie got his PhD, and the scientific world had to come to terms with the fact that just as light, which they were used to thinking of as a wave, could also be described in terms of particles, so the electron, which they were used to thinking of as a particle, could also be described in terms of waves. In 1927, both an American team of physicists and George Thomson in England carried out experiments demonstrating the wave behaviour of electrons, scattering them from crystals. The wavelengths (frequencies) of electrons with a certain energy, measured in this way, exactly match Planck's formula $E = hf$. George Thomson, who thereby proved that electrons are waves, was the son of J. J. Thomson, who, a generation before, had first proved the existence of electrons as particles.

The notion of 'wave–particle duality' became one of the key ingredients in the quantum theory that was developed in the mid-1920s, and which Richard Feynman studied as an undergraduate. In fact, the quantum theory was developed twice at that time, almost simultaneously, once using what was essentially a particle approach and once using what was essentially a wave approach. The leading light in the development of the particle version was Werner Heisenberg, the first major participant in the quantum game to have been born in the twentieth century (on 5 December 1901, at Würzburg, in Germany). A variation on this theme (in many ways, more complete) was also developed independently by another young physicist, Paul Dirac, who was just a few months younger than Heisenberg, having been born at Bristol, in England, on 8 August 1902.

Erwin Schrödinger, an Austrian physicist, was the odd one out among the pioneers of the new quantum theory, having been born in 1887, and had obtained his doctorate back in 1910. He built from De Broglie's ideas about electron waves, and came up with a version of quantum theory that was intended to do away with all the mysterious jumping of electrons from one level in an atom to another, deliberately harking back to the classical ideas of wave theory.

It was Dirac who proved that all of these ideas were, in fact, equivalent to one another, and that even Schrödinger's version did include this 'quantum jumping', among other things, in its equations. Schrödinger was disgusted, and famously commented of the theory he had helped to develop, 'I don't like it, and I wish I'd never had anything to do with it.' Ironically, because most physicists learn about wave equations very early in their education, and feel comfortable with them, ever since quantum mechanics was established in the 1920s it is Schrödinger's version that has been most widely used for tackling practical problems, like interpreting spectra.

We don't want to go over the whole story of the development of quantum theory in the 1920s here,[4] and instead we'll jump straight to the final picture, which can best be understood (as far as anything in quantum physics can be understood) in terms of an example which Feynman himself would, much later, call the 'central mystery' of quantum mechanics. It is the famous 'experiment with two holes'.

In this example you can imagine sending either a beam of light or a stream of electrons through two tiny holes in a screen – the experiment has actually been done with both, and everything we are going to discuss here has been proved by experiment. When waves travel through two holes in this way, the ripples fan out from each hole on the other side of the screen and combine to form what is called an interference pattern, exactly like the interference pattern you would see on the surface of a still pond if you dropped two pebbles into it at the same time. In the case of light, this basic experiment was one of the techniques used to prove, early in the nineteenth century, that light is a wave – a second screen placed beyond the one with the two holes will show a pattern of light and dark stripes, 'interference fringes', produced in this way (see Figure 4*a*).

But if individual particles (such as electrons) are fired, one at a time, through the experiment with two holes, you would expect, from everyday experience, that they would pile up in two heaps, one behind each of the holes. A suitable detector screen on the other side (essentially the same as a TV screen) ought, if electrons are particles, to show two blobs, corresponding to the trajectories of electrons going through either of the two holes. But it doesn't. Here's what happens. Each individual particle starts out on one side, passes through the experiment, and strikes the detector screen. Surely, you would think, each particle can only go through one hole or the other. And, to be sure, each particle makes just one spark of

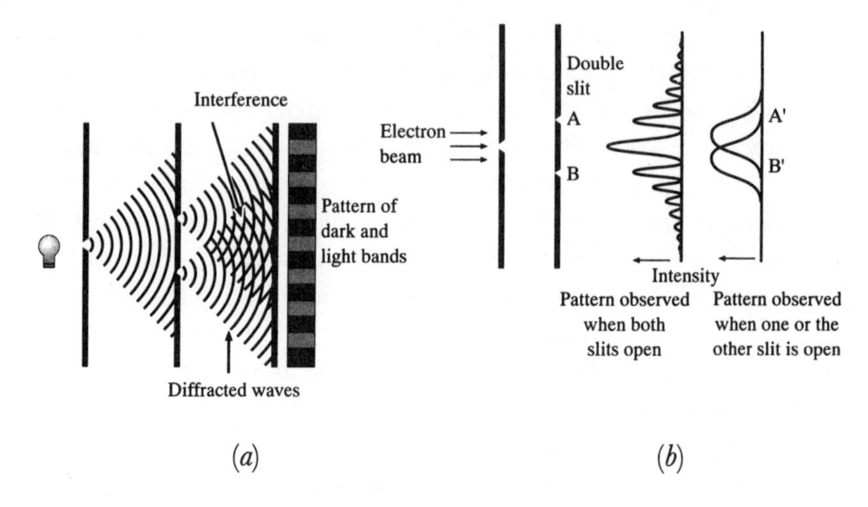

Figure 4. (*a*) When light spreads out from a pinhole in a screen to pass through two pinholes in a second screen, the pattern made up by the light from the second two holes shows alternating dark and light bands, exactly as if the light is behaving as waves which interfere with one another. (*b*) When electrons (or, indeed, photons) are fired through a similar set-up with one hole open, they pile up, like particles, in one heap behind the open hole. But if both holes are open, the 'particles' somehow interfere with each other to produce a pattern exactly equivalent to the pattern produced when waves interfere. This is the central mystery of quantum mechanics, the experiment with two holes. How do the electrons know in advance whether one or both holes are open, and adjust their behaviour accordingly?

light on the detector screen, indicating that it arrives as a particle. But after thousands of particles have been fired through the experiment one after the other, a pattern of sparks of light builds up on the detector screen. Not the two blobs behind the two holes that you might expect from your everyday experience of how particles behave, but the familiar interference pattern for waves! (See Figure 4*b*.) We stress that this experiment really has been done, and both electrons and photons behave in this way. It is as if each particle flies through both holes at once, interferes with itself, decides where it belongs in the interference pattern, and goes there to make its own individual contribution to the emerging pattern. Quantum entities seem to travel as waves, but to arrive (and depart) as particles.

As well as wave–particle duality, this example highlights another aspect of the quantum world – the role of probability. Nothing is certain in the quantum world. For example, before an individual electron is fired through the experiment with two holes it is impossible for the experimenter to say exactly where on the screen on the other side it will arrive. You can only calculate, in accordance with the rules of quantum probability, the chance of it ending up in a particular part of the interference pattern. It is likely to turn up in one of the bright parts of the pattern and unlikely to appear in one of the dark stripes in the pattern, but that is all you can say. Quantum processes obey the same rules of chance as the dice at a craps table in Las Vegas, which prompted Einstein to express his own disgust with the theory with his comment, 'I cannot believe that God plays dice.'

So how should we think of an electron 'in' an atom, where it is 'travelling' in its 'orbit', rather than 'arriving' at a detector? The standard picture, used by physicists for the past 70 years, says that the electron cannot be located at any one point in space near the nucleus, but that the location of each electron is spread out over a region of space surrounding the nucleus – not just stretched out along a single orbit (like the orbits of the planets around the Sun), but spread out in a shell literally surrounding the nucleus, a shell called an 'orbital'. The orbital is thought of as a 'cloud of probability', representing the likelihood of finding the electron. If a measurement were carried out that was accurate enough to locate the precise position of the electron, for that instant it would indeed 'arrive' at some definite position within the orbital, and manifest itself as a particle. The position it would arrive at would be subject entirely to chance, selected at random from the options open to it. But as soon as the observation had been completed, the electron would dissolve once again into a haze of probabilities. And this kind of behaviour is supposed to represent the behaviour of *all* quantum entities.

The way in which an entity such as an electron manifests itself as a particle when it is measured is called the 'collapse of the wave function', and all quantum systems are supposed to exist in some sort of state of probabilistic uncertainty until an observation or measurement is made and the wave function collapses.

This has given rise to all kinds of debate about what constitutes a measurement, and when exactly the wave function collapses, which, happily, we do not need to go into here.[5] It sounds bizarre. And yet, it works. Quantum theory says that at the level of atoms and subatomic 'particles', entities have to be thought of as having properties of both wave and particles, that nothing is certain, and the outcome of an experiment depends on chance, in the strict mathematical sense. But all of this strange mixture has practical applications. Since the face shown by an atom to the world – to other atoms – is its electron cloud, and chemistry depends on the way the electron clouds of different atoms interact with one another, it is this quantum mechanical view of the behaviour of electrons that underpins, among other things, the extremely successful modern understanding of chemistry, developed in the wake of these discoveries.[6]

In spite of its weirdness, the new quantum mechanics worked. The point was made, in forceful terms, by the greatest single triumph of this period, when, in 1928, Dirac published an equation which incorporated the requirements of the Special Theory of Relativity into the quantum theory to provide a complete description of the electron, in terms of relativistic quantum mechanics. The Dirac equation described everything there was to know about the electron, and made predictions which matched the results of all the experiments. It also made another prediction, which even Dirac did not immediately interpret correctly, and which would have a profound influence on the career of Richard Feynman, who was just 10 years old when Dirac came up with his equation.

Dirac's equation not only explained everything there was to explain about an electron, but it did so in duplicate. The point is, there were two sets of solutions to the equation. Now, there is nothing unusual about this. If you see an equation such as $x^2 = 4$, you know that the solution to the equation is $x = 2$, because $2 \times 2 = 4$. But there is, in fact another solution as well. Because two negative numbers multiplied together make a positive number (just as in language, where a 'double negative' makes an affirmative), $(-2) \times (-2) = 4$, as well. So -2 is a perfectly good solution to the equation $x^2 = 4$, if you are a mathematician. Such 'negative roots' often crop up in equations, and the question is whether they mean anything in practical terms. The second solution to Dirac's equation describes particles identical to electrons but with negative energy. Most people would probably have dismissed this as a meaningless mathematical quirk. But Dirac's genius led him to wonder 'what if' – what if these negative-energy electrons really existed?

The big snag was that if you allow electrons to have negative energy, at first sight it seems that they *all* ought to have negative energy. Like water running downhill, any physical system seeks out its lowest possible energy level. If there were 'negative-energy levels' for electrons, then obviously even the highest of these levels would be below the lowest positive-energy level, and all electrons would fall down into the negative levels, radiating a blaze of electromagnetic energy as they did so. But suppose, Dirac argued, all of the negative-energy levels were full up, just as the sea is full up with water. Water running downhill would carry on running down to the bottom of what is now the sea, if there were no sea in the way; but in the real world rivers only run down to discharge their water into the top of the sea, because the sea is already full up. If all of the negative-energy 'sea' were full of electrons, the only openings available for any more electrons would be the positive-energy levels above. The negative-energy electron sea would be completely undetectable, or invisible, because it was the same everywhere.

But now Dirac went a step further. In the everyday world, an object in a low energy state can be kicked up to a higher energy state by an input of energy – literally kicked, perhaps, like a ball being kicked up a flight of stairs. What if the negative-energy electron sea were *not* quite the same everywhere? Suppose an energetic interaction of some kind – perhaps the arrival of a cosmic ray from space – gave energy to one of the invisible electrons in the negative-energy sea, and kicked it up into a state with positive energy? Now, the electron would be detectable ('visible') to physicists as a normal electron. But it would have left behind a 'hole' in the negative-energy sea. Electrons have negative electrical charge, so, as Dirac pointed out at the end of the 1920s, the hole in a sea of negative charge would behave exactly like a particle with positive charge (absence of negative being the same as presence of positive). If the hole were near a detectable visible electron, for example, negative-energy electrons in the sea would be repelled from the visible electron and would try to escape by hopping in turn into the hole; as one neighbouring invisible electron hopped in, the hole would fill up, leaving a hole where that invisible electron had been, and so on. The effect would be that the hole would move towards the visible electron, behaving just like a positively charged particle. To see what happens when the hole meets the visible electron, read on.

At this point, Dirac had a failure of nerve. Taking his equation at face value, the only physical meaning you could reasonably give to the hole would be as a particle exactly like the electron except for its positive charge. But in 1928, remember, physicists only knew two kinds of particle, the electron (with negative charge) and the proton (much more massive, but with a positive charge the same size as the electron's negative charge). Even the neutron had not yet been discovered. So Dirac suggested in his paper that the holes in the negative-energy electron sea could be identified with protons. This really didn't make sense, and partly as a result nobody really quite knew what to make of the notion of the negative-energy electron sea and its holes at first. But then, in 1932 the American Carl Anderson discovered traces of particles which behaved exactly like electrons but with positive charge, in cosmic ray experiments (cosmic rays are particles that arrive at the Earth from space). He concluded that the 'new' kind of particle was a positively charged counterpart to the electron, and gave it the name positron (an example of what is known as antimatter); it had exactly the right properties to match the behaviour of Dirac's holes. The same year, James Chadwick, in Britain, identified the neutron.

Almost overnight, the number of kinds of individual particles known to physicists had doubled, from two to four, and their view of the physical world was transformed. You can get an idea of the dramatic impact of these discoveries on the physics community by the speed with which the Nobel committee responded to them. In 1933, Dirac received the Nobel Prize in Physics (he deserved it anyway, but the successful 'prediction' of positrons clinched it); in 1934, there was no award (an astonishing decision to modern eyes!); in 1935 it was Chadwick's turn; and in 1936 Anderson received the prize.

Since then, a wealth of other subatomic particles have been discovered, and each variety has its own antimatter counterpart. All of this can be explained by variations on the hole theory, and that theory does still provide one of the best mental pictures of how energy is liberated when a particle (such as an electron) meets its antiparticle counterpart (in this case a positron) and annihilates, leaving nothing behind but a puff of energy. The electron has fallen into the positron hole, releasing energy as it does so, and both hole and electron simply disappear from the everyday world, cancelling each other out. Or, if energy is available (perhaps from an energetic photon) a negative-energy invisible electron can be kicked out of its hole and promoted into visibility, creating, along with the hole it left behind, an electron–positron pair.

But although the physical picture is simple and rather appealing (if you can live with the idea of a sea of negative-energy invisible electrons), the mathematics of the hole theory turned out to be rather cumbersome as a means of describing particle interactions. By the time Dirac received his Nobel Prize, the person who would demonstrate a much simpler way of describing interactions involving electrons and protons was just starting his final year in high school in Far Rockaway. Even though details of all the new discoveries had not yet filtered down into the standard textbooks and courses taught at universities (not even at MIT), Richard Feynman was exactly of the generation to be brought up on the new physics as an undergraduate, and to be prepared to carry things a stage (or two) further when it became time for him to make his own contributions to science. It helped, of course, to be a genius. A genius like Feynman would have made a mark on science whenever he had been born; but the accident of the timing of his birth decided the kind of mark he would make. As a member of the first generation to be brought up on quantum mechanics, he carried the triumphant, but still incomplete, theory through to its greatest fruition.

Even though the standard undergraduate textbooks might not yet tell the full story of quantum mechanics, Dirac himself had written a definitive account in his book *The Principles of Quantum Mechanics*, first published by Oxford University Press in 1930, which was the first comprehensive textbook on the subject. It came out in a new edition the year that Feynman set out for MIT, and the book (which later went through further revisions) is still the best introduction for serious scientists. The 1935 edition would have a profound influence on the young physicist at MIT – but at the time he started his undergraduate courses there, he didn't even know that he was a physicist.

Notes

1. The full story of the development of Einstein's ideas is told in *Einstein: A Life in Science*, by Michael White & John Gribbin (Simon & Schuster, London, 1993; Dutton, New York, 1994).
2. James Clerk Maxwell, *A Dynamical Theory of the Electromagnetic Field*, 1864; see, for example, Ralph Baierlein, *Newton to Einstein* (Cambridge University Press, 1992), p. 122.
3. See note 1.
4. If you do want the details, see John Gribbin, *In Search of Schrödinger's Cat* (Bantam, New York & London, 1984).
5. But see *Schrödinger's Kittens*.
6. Largely by the American Linus Pauling, who summed up the work in his book *The Nature of the Chemical Bond* (Cornell University Press) in 1939, and received the Nobel Prize for his work in 1954.

— 3 —
College boy

New students at MIT had to find a fraternity which they could join, to provide them with a home and a social group within which they would fit into the college community. This system was basically a good one, in which senior students would look after freshmen in their own fraternity, teaching them the college ropes and looking out for their interests; occasionally, rivalry between fraternities and ragging of younger students by older ones got out of hand, but this doesn't seem to have been a problem at MIT in Feynman's time there.

For many students, the process of joining a fraternity would involve offering themselves to different fraternities, and trying to persuade them that you were a desirable prospective member of the group. For the best students, like Feynman, it worked the other way around. The fraternities sought you out, and tried to persuade you to join them. In fact, in Feynman's case the choice (or competition) was limited. There were only two Jewish fraternities at MIT, and there was no way, in those days, that Richard could join a non-Jewish fraternity. This 'Jewishness' had nothing to do with religion, which Feynman had long since abandoned; it simply had to do with your family background. Both these fraternities were on the lookout for bright students, and held gatherings called 'smokers' to get to know boys from New York who were going to MIT.

Feynman, who still thought of himself as a mathematician at this time, went to both these smokers. At one, for the fraternity Phi Beta Delta, he discussed science and maths with two older students, who told him that since he knew so much maths already he could take examinations at MIT as soon as he arrived there, which would allow him to skip the first-year course and go straight on to the second-year work in the subject. Both Phi Beta Delta and the rival fraternity, Sigma Alpha Mu, were eager to enrol Feynman, who was obviously going to be the kind of student that would add lustre to their groups (but don't run away with the idea that fraternities were only interested in academic ability; they were just as eager to attract students with other talents, such as sportsmen). Partly on the strength of the good advice he had already received from them, Feynman agreed to join Phi Beta Delta.

When the time came to leave Far Rockaway for MIT, however, some of the students from Sigma Alpha Mu called round. They would be driving up to college, and offered Feynman a ride, which he happily accepted. Like all mothers in such circumstances, Lucille watched with mixed feelings when the day came and, as arranged, her son drove off with a bunch of strangers on the journey to Boston, on what became a snowy day with tricky driving conditions. But Feynman was elated that he was being treated like an adult: 'it was a big deal; you are grown up!'[1]

But the deal wasn't quite that simple. What Sigma Alpha Mu had done, in effect, was to kidnap Feynman, hoping to enrol him with their fraternity before their rivals at Phi Beta Delta realized what was going on. They suggested, having arrived late in Boston, that he stay the night in their house, and he agreed, not realizing that he was the subject of this tug of war. In the morning, two of the seniors from Phi Beta Delta turned up to claim their own, and after some discussion Feynman finally did become a pledge at Phi Beta Delta, feeling a warm glow at being the centre of all this attention; partly as a result, he immediately began to overcome his old self-consciousness about being a sissy that everybody laughed at. The other fraternity members soon helped to develop his social skills further, although he never became what you would call a conformist in social matters.

Just before Feynman had joined Phi Beta Delta, the fraternity had almost collapsed because of a conflict of interest between its members. About half the fraternity brothers were wild socialites, who had cars, knew all about girls and organized dances. The rest were serious academic types, who studied all the time, were socially gauche and never went to the dances. In *Surely You're Joking*, Feynman recounts how, in order to avoid breaking apart entirely, the fraternity members had got together and agreed to help each other. Everybody in the fraternity had to achieve a certain grade level in their courses, and if one of the socialites was having difficulty then the academics were obliged to help them get up to the required standard. In return, everybody, including the academics, had to go to every dance. The socialites would help by teaching the others how to dance and other social niceties, and even by making sure that they each had a date for the evening. Apparently, the system worked beautifully, and was an ideal way for Feynman to learn how to socialize. 'It was', he said, 'a good balancing act.'

Not that he didn't have some difficulty with the lessons. It had to be explained to him, for example, that it was not done to invite waitresses to the dance, and although he still lacked the confidence to disregard their advice in this regard, he couldn't resist teasing his new friends with displays of the stereotypical Brooklyn character that he later played to perfection. On one occasion, a date was arranged for him with a girl called Pearl. Before he met her, he made a great play of pronouncing her name 'Poil' to his peers, who were horrified that he would let them down on the big occasion. So when he met the girl – pronouncing her name perfectly, of course – he explained to her that he was going to have a little joke at the expense of the other fraternity members, and spent the whole evening introducing her to friends at the dance as 'my goil, Poil'.

But as well as confirming his fondness for pranks and eagerness to puncture pomposity, the story highlights another facet of Feynman – the way he was able to charm people, especially women, into going along with him even if, as in the case of 'Poil', they had only just met. It is easy to imagine the average guy, even if he had dreamed up this prank, being met with a distinctly frosty reception when he explained it to the girl. But not Dick, as he was now becoming known to his friends. He was tall enough, at just under six foot, and certainly dark and handsome, devastatingly attractive, amusing, and, when he wanted to be, charming.

If nothing else, the kind of behaviour typified by the 'Poil' incident helped to ensure that from the moment he arrived at MIT, Feynman was never known as a sissy. He made doubly sure on an occasion early in his time there when a gang of sophomores raided the Phi Beta Delta house intending to tie the freshmen up, take them out into the woods and dump them there for a long walk home.* In order not to look like a sissy, instead of going along quietly Feynman fought so vigorously that it took several of the older students to subdue him, and he gained an immediate reputation for being a tough guy that nobody ought to mess with. The exaggerated Brooklynese was all that he needed, after that, to maintain the reputation.

Throughout his time at MIT, though, Feynman's real 'goil' was still Arline. By mutual agreement, he went out with other girls when she wasn't around, and she went out with other guys, but they wrote to each other and Arline became almost part of the Feynman family, visiting the house in Far Rockaway to give piano lessons to Joan, painting with Melville, and going to cookery classes with Lucille. She also visited MIT occasionally, and they saw each other during the vacations – it was during the midwinter break of Feynman's freshman year that they agreed to marry when he had finished his studies, and from then on they regarded themselves as engaged.

But the romance was not all plain sailing, even before Arline became ill. One summer, Richard stayed in Boston, working in a summer job for Chrysler, investigating friction. Arline had arranged a job in Scituate, about 20 miles away, in order to be near him, but was talked out of it by Melville Feynman. In spite of her now being a family friend, as well as Richard's girl, Melville seems to have been concerned that she might have an adverse influence on Richard's career. In those days, marriage was out of the question for a student, and Melville had put everything he had – not just money, but a major emotional investment – into giving his son a chance to become a real scientist. He wasn't going to let *anything* stand in the way of that. Nevertheless, the couple still got together a few times during that summer.

By then, Richard was no longer a mathematician. Some time during his first year at MIT, Feynman had begun to ask himself

*For those who are not familiar with the American educational system, a freshman is a first-year undergraduate, a sophomore is a second-year student, a junior is a third-year student on a four-year course, and a senior is a final-year undergraduate.

what mathematics was really useful for, and decided that the only thing you could do in the way of a career in maths was teach it to somebody else. In an initial over-reaction, seeking something more practical, he switched his major at first to electrical engineering; but then he realized that he had gone too far, and switched again into the middle ground as a physics major. That gave him the opportunity for the 'hands on' laboratory work that he loved (one of his favourite undergraduate experiments involved measuring the speed of light), but also gave free rein for his more abstract thinking about the nature of things. But whatever the course he was officially signed up for, throughout his time at MIT Feynman continued to learn more science from books and discussions with other bright students than from the standard undergraduate courses. He also benefited from the flexibility of MIT in allowing any student who was bright enough to take any course on offer, no matter how advanced it was supposed to be.

In his freshman year – when he was already, remember, doing the sophomore course in mathematics – Feynman's roommates at the fraternity house were two senior students, Art Cohen and Bill Crossman. They were taking an advanced course in physics, intended for seniors and graduate students, which had recently been devised by John Slater. It was based on his own book *Introduction to Theoretical Physics*. Slater was the head of the physics department at MIT, and had worked in Europe, where he had learned first hand about the new quantum mechanics; the course didn't quite go that far, but it did introduce the new atomic theory and the wave concepts that were becoming so important in quantum physics. Unlike some of his contemporaries, however, Slater didn't worry about the seemingly mystical aspects of quantum theory – the way in which an entity could be both particle and wave, or the way in which a photon seemed to know in advance about the set-up of the experiment with two holes before it passed through the apparatus. He was a pragmatist, who asked only that theories should be able to predict the outcome of experiments with reasonable accuracy, a philosophy that he tried to pass on to his students. Just how the photon got from A to B didn't matter, as long as the theory could tell you that it would indeed, if it started out from A in a certain way, end up at B.

Feynman used to listen to Cohen and Crossman discussing problems they had been set in Slater's course. After a couple of months, he was confident enough to chip in when they were worrying about how to solve some problem. 'Hey,' he said, 'why don't you try Baronally's equation?' Cohen and Crossman had never heard of 'Baronally'. The trouble was, being self-taught and only ever having seen the name in a book, he had hopelessly mispronounced the name 'Bernoulli'. But eventually communication was established. They tried the equation, and it worked. From then on, the two seniors were always ready to discuss their physics problems with Feynman, and although he couldn't do them all, he often knew some trick, like Bernoulli's equation, that would set them on the right trail. And by talking about the problems, of course, he picked up a lot more so-called advanced physics. By the end of the year, he had decided that he knew enough to tackle the course (aimed, remember, at seniors and graduate students) in his sophomore year.[2]

When he turned up to register for the course, Feynman was wearing his Reserve Officer Training Corps (ROTC) uniform, which was compulsory for first- and second-year students. All the seniors and graduate students wore their everyday clothes. They had green or brown cards to fill in to register, corresponding to their status; Feynman had a pink card. In addition, he looked even younger than he really was. It all made him feel good; he liked to be seen as the boy genius. This time, though, he wasn't alone. Another student in ROTC uniform, carrying a pink card, came and sat next to him. It was another boy genius, another sophomore, Ted Welton, who also had enough self-confidence to sign up for the advanced course.

The two prodigies cautiously got to know each other, verbally circling around one another to see if they would be rivals or friends. Feynman noticed that Welton was carrying a book on differential calculus that he had wanted to get out of the library. Welton discovered that a book he had been trying to find in the library had been taken out by Feynman. Feynman claimed he had taught himself quantum mechanics already, using Dirac's book; Welton claimed that he had learned all about the General Theory of Relativity. Each was impressed by the other. They decided that 'cooperation in the struggle against a crew of aggressive-looking seniors and graduate students might be mutually beneficial',[3] and soon became firm friends.

Even among the aggressive-looking seniors and graduate students, Feynman stood out. For the first semester, the course was taught by Julius Stratton, a young physicist who certainly knew his stuff (he went on to become President of MIT) but sometimes didn't prepare his presentation with due care and attention. Whenever he got stuck in the middle of a lecture, he would turn to the audience and ask, 'Mr Feynman, how did you handle this problem?', and Dick would take over. 'I note', Welton recalled many years later, 'that Stratton never entrusted his lecture to me or to any other student.'[4]

Quantum mechanics appeared formally in the second semester of the course, and was taught by another young physicist, Philip Morse. Feynman and Welton, having worked together through some introductory texts by then, swallowed this up and were eager for more. They asked Morse where they could go for the real quantum nitty gritty, and as a result he invited them, during their junior year, to visit him one afternoon a week, along with a promising student in his senior year, for special tuition in the subject. Eventually Morse gave them real problems to solve using quantum mechanics – such as the separation of the energy levels for the electron in a hydrogen atom. This brought home to them, forcefully, that it wasn't just some abstract theory, but practical science which could indeed be used to solve real problems.

Feynman also swallowed up courses in chemistry, metallurgy, experimental physics and optics – anything scientific was meat and drink to him. When a new course in theoretical nuclear physics, intended for graduate students, was offered for the first time at MIT, he went along to sign up for that as well. There was a crowd of students already in the room, and Morse was sitting on the window sill. He looked up, and asked if Feynman intended to register for the course. Feynman replied that he did. Morse asked if Welton was coming along. Feynman said yes. Good, said Morse; that meant they could start. It turned out that the rules required at least three students to enrol formally on the course, for credit, before it could be given. Only one of the graduate students had been willing to sign up for it. The others were afraid that they might flunk it, damaging their grade averages, but were eager to sit in on the course as observers, without being examined on the subject, if it did take place. So two of the three officially enrolled students for the new graduate course were actually undergraduates. Feynman, in the end, found it all quite straightforward – and passed the graduate course with flying colours.

There was one outstanding oddity about the way Feynman did his science at MIT, in the light of how he later made his mark in science. He liked to solve problems 'properly', by working out the relevant equations – in the case of a ball flying through the air, for example, this would involve solving Newton's equations of motion. There was an easier way, which the students at MIT were taught, called the Lagrangian approach, after the French mathematician Joseph Louis Lagrange, who lived from 1736 to 1813 and was made a Count by Napoleon Bonaparte. The beauty of the Lagrangian approach is that it doesn't involve calculating the changing forces and accelerations affecting, in this example, the flight of a moving object instant by instant, but deals only with the overall energies involved and the elapsed time.

Sound familiar? The Lagrangian approach is, indeed, directly based on the Principle of Least Action, that Feynman had fallen in love with when it had been introduced to him by Bader in high school. Why he eschewed this approach as an undergraduate remains a mystery, but the most likely explanation is his love of both problem solving (preferably from first principles) and showing off. While his fellow students, including Welton, were solving the problems the easy way, using the Lagrangian, Feynman would solve them even more quickly (in almost all cases) the hard way, integrating the equations of motion as laid down by Newton – a technique often known as the Hamiltonian method, after the nineteenth-century Irish mathematician William Hamilton. This involved working with 'the Hamiltonian', an appropriate set of differential equations describing the system being investigated.

'My way would take ingenuity,' Feynman later said,[5] 'whereas the trick of the Lagrangian was that you could do it blindfold.' Shades of the old days of the Interscholastic Algebra League! The trouble was, for the problems the students were given at undergraduate level, the Lagrangian approach was simply too easy for

Feynman to bother with; it hardly gave him scope to exercise his brain. But he learned the approach anyway, if only to be able to test it against the conventional methods, such as the Hamiltonian approach, and see which was really quickest in a variety of situations. And in a few years' time, when he came up against some *really* tricky problems, he was happy to use the technique to solve them.

But if Feynman found the science taught at MIT so easy that he had to make his own difficulties in order to make the problem solving more interesting, outside the sciences it was a different matter. In a letter to a friend soon after he started at MIT, Feynman described the courses he was taking as 'physics, math, chemistry, ROTC, English; in decreasing order of pleasure I get out of them'.[6] But he soon found something even worse than English that he had to struggle with in order to keep up his overall grades and be allowed to graduate.

MIT quite rightly required all their students to take (and pass) three humanities courses, in order to become at least slightly more well rounded as citizens by the time they graduated. English, like it or not, was compulsory, but to Feynman's delight he found astronomy listed as a humanities course, so that was no problem. But for his third choice, after rejecting possibilities such as French literature, he was left with philosophy, which at least sounded as if it ought to have some bearing on science. But he was wrong, at least as far as the philosophy being taught to undergraduates at MIT in the 1930s was concerned.

In *Surely You're Joking*, he explained how he scraped through the English and philosophy courses without bringing shame to the fraternity – for, of course, in these cases the boot was on the other foot, and it was Feynman who was obliged to seek help and advice from the others in order to achieve the standard that the fraternity felt was acceptable for one of its members.

In English, for example, on one occasion the assignment was to write a theme on Goethe's *Faust*. Feynman was in despair, unable to come up with anything, and threatening not to hand in any work at all. His fraternity brothers persuaded him that he had to write something – anything – just to prove that he wasn't trying to get out of doing the work. So he wrote an essay on the theme 'On the Limitations of Reason', discussing the relevance of moral values, scientific methods of reasoning and so on. But there was nothing about *Faust.* One of the fraternity brothers read the theme, and advised Feynman that what he should now do was add a few lines linking what he had said to *Faust.* It seemed ridiculous, but under pressure from his peers Feynman complied, adding half a page saying that Mephistopheles represents reason, Faust the spiritual, and that Goethe's aim in writing *Faust* was, indeed, to show the limitations of reason.

The professor was completely taken in. He commented that the introductory material was good, even if the direct references to *Faust* were rather brief, and awarded Feynman a B^+. More confirmation that English was a 'dippy' subject – but the grade was up to the requirements demanded by the fraternity.

Philosophy, though, was beyond mere dippiness. According to Feynman, the professor who gave those classes, an old man with a beard, mumbled so much that Dick could not understand a word he was saying. To pass the time in class, Feynman used to drill holes in the sole of his shoe, using a one-sixteenth drill bit that he carried in his pocket, twisting it between his fingers. The crunch came when it was time to write a theme at the end of the course. The only words that Feynman could recall from the weeks of lectures were 'stream of consciousness'. That gave him the idea of writing about what happens to the stream of consciousness when you go to sleep – how does it switch off?

Formulated that way, the project became a scientific experiment. There were four weeks to go before the theme had to be handed in, and every afternoon (as well, of course, as every night) Feynman would go to his room, lie down and go to sleep, while trying to observe mentally what was happening. He noticed, among other things, that as he dropped off to sleep the flow of ideas still seemed to his consciousness to be logically connected, even as they became more jumbled. He watched his mind 'turning off', and wrote a theme about his experiences. To round it off, he ended with a little verse:

> I wonder why. I wonder why.
> I wonder why I wonder.
> I wonder *why* I wonder why
> I wonder why I wonder!

At the end of the course, instead of a final lecture the professor picked out a few of the better themes to read to the class. To Feynman, sitting twisting his drill bit into the sole of his shoe, it was the same old gibberish. As far as he could tell, the professor was mumbling something along the lines of 'Mum bum wugga mum bum . . . ' Dick had no idea what the theme was about. The professor came to another theme, and read on: 'Mugga wugga mum bum wugga wugga . . . ' Neither Dick nor his drill bit had a clue what it was about, until the professor got to the end, and recited:

> Uh wugga wuh. Uh wugga wuh.
> Uh wugga wugga wugga.
> Uh wugga *wuh* uh wugga wuh
> Uh wugga wugga wugga.

It was only then that Feynman realized his contribution had been singled out for praise.[7] He got an A for his theme, without having understood anything the professor had tried to teach during the course. In English, he had at least been aware of the plot of *Faust*, and made some effort to mention it in his theme. His belief that philosophy was completely idiotic was reinforced, but again he had achieved a good enough grade.

In fact, during his time at MIT Feynman didn't do anything outside science unless he had to. ROTC was compulsory, so he joined; but he didn't join any other clubs or societies. The fraternity dances were compulsory, so he went along, and benefited enormously from the experience; but otherwise his idea of a wild time was discussing physics with Welton. He was aware of the financial pressures on his family, and earned what money he could to help out. But he didn't work behind the counter in a drug store, or pumping gas; he worked as an assistant doing various odd jobs for the professors at MIT, and in summer jobs with a scientific flavour. Somewhere along the line, though, one of his lifetime hobbies, drumming, got started while he was at MIT (the other hobby, art, got started much later). He banged on walls, tables, pots and pans – anything he could use to beat a rhythm – and enjoyed listening to African drum music, although he never enjoyed 'ordinary' music, and described himself as being tone deaf.

Feynman's scientific achievements as an undergraduate, especially in his senior year, were so good that he had two scientific papers published in the *Physical Review* before he even graduated (more about those pieces of work in Chapter 4). He liked MIT, and wanted to stay on there to do research, working for his PhD. It was the only scientific world he knew, and he thought it must be the best school in the country – if not the entire world – to do science. But Slater, who knew the high-flying student well by then, wouldn't permit it. He told Feynman that he had to go to another school to complete his education, and Feynman was later grateful for the advice. 'Slater was right. I learned that the world is bigger and there are many good places.'[8] The 'good place' that Feynman went to, after he graduated from MIT in 1939, was Princeton.

As early as January 1939, Slater and Morse had advised their colleagues at Princeton that something special was coming up. The advice was necessary, because Feynman's academic record was a bizarre mixture of the nearly perfect and the truly dreadful. John Wheeler, who became Feynman's thesis adviser at Princeton, has told how baffled the Graduate Admissions Committee at Princeton was by Feynman's scores in the standard aptitude tests.[9] In physics, he was literally perfect – 100 per cent. The maths score was nearly as good – both were the best the Committee had ever seen. But they had never admitted anyone with such low scores in history and English (it doesn't bear thinking how low those scores would have been without all that help from his fraternity brothers). What tipped the balance was his practical experience in chemistry and investigating friction. The odd jobs for the professors at MIT, and the summer job for Chrysler, paid a dividend that Feynman can never have anticipated, and he was duly admitted to Princeton in the autumn of 1939.

There had been one other, almost unmentionable, hurdle to overcome. Princeton didn't actually have a formal Jewish quota, but on the other hand they didn't want the place overrun with Jews. The head of physics at Princeton, H. D. Smyth, made delicate inquiries of MIT; Slater and Morse replied that although Feynman was Jewish, he didn't look it and had an attractive personality, as well as being the best student they had seen for many years.[10] 'I guarantee you'll like him', Slater told Smyth. Morse was equally enthusiastic, describing Feynman as 'a pleasure to work with. One only needs to give him a few suggestions to keep him going on research; and his abilities make him capable of covering a large amount of territory in a short time.' With recommendations like that, the Jewish background was never really going to be a problem, even in the culture of 1939.

Someone else, though, was worried about the problems of Jews finding employment in the United States as the 1930s gave way to the 1940s. In his autobiography,[11] Morse mentions a visit Melville made to see him around the time Dick was graduating. Having explained that the family could just barely afford to finance Richard through another four years of school, Melville asked for reassurance that the effort would be worth it. Was Richard good enough? Jobs in physics were hard to get in 1939, and the unspoken question behind Melville's inquiry was whether they would be impossible to get for a young physicist from a Jewish background. Morse writes in his autobiography that he reassured Melville, telling him that Richard definitely was good enough to justify the continuing investment in his education. But there is more to the story than this.

Joan Feynman explains that, with hindsight, it became clear that the main reason for Melville's concern was that he had health problems, and knew that his high blood pressure meant that he would not live long – probably not long enough to see Joan through college (in fact, he died when she was in her freshman year, by which time his annual income had passed $10,000; from then on she funded her education with scholarships and other aid). The family was already saving to provide for Lucille and for Joan's education when and if the inevitable happened, so Melville's concern about Richard's prospects was well founded.[12] Nevertheless, it is a sad reflection on the almost automatic antisemitism of the times that this should have been an additional cause for concern for Melville at an already difficult time.

The Admissions Committee at Princeton was, in the end, sufficiently impressed by Feynman not just to admit him, but to offer him a research assistantship, which meant that he actually got paid for helping a more senior scientist with his research and his undergraduate teaching, while working for his own PhD as well. This must have been a great relief to Melville. The scientist Feynman was assigned to was Wheeler. He was 28, and Feynman 21, when they met for the first time. Perhaps over-conscious of his own relative youth, Wheeler (who was a first-rate scientist and had already worked for a couple of years with Niels Bohr's group in Copenhagen) tried to establish what he regarded as the proper professor–student relationship from the outset.

The full flavour of the first encounter between Feynman and Wheeler doesn't always come across in books about him, but it was a highly significant meeting of minds that set the scene for a fruitful collaboration between two scientists who were both open to new ideas in physics, no matter how wild. It was obvious to anyone who knew him, even slightly, that Feynman had this crazy kind of genius. But Wheeler has always seemed, from the outside, to be a much more sober kind of person. He wears suits and ties, he is calm and respectable, he doesn't play the bongo drums or crack safes. But behind the façade lies one of the best ideas brains of the past 60 years, an expert on exotica such as black holes (he coined the term in its astronomical sense) and parallel realities. Reading some of Wheeler's scientific papers, it is hard to believe that the bizarre images they conjure up spring from the mind of a man who looks like the head of an old-fashioned bank.

As a pompous and somewhat self-important 28-year-old, though, who had yet to make his own mark on science, Wheeler felt that his time was too valuable to squander overmuch on new graduate students. He made an appointment to see Feynman at certain times every week, and told him that each meeting would last for a certain time. It is easy to imagine the freewheeling Feynman's internal reaction to this rigid timetabling. At the start of the first of these formal meetings, Wheeler made a show of pulling out his expensive pocket watch and placing it on the table, so that he would know when Feynman's time was up – and so that Feynman would know his place in the pecking order. Well, thought Dick, two can play at that game. Before the next meeting, he bought a cheap pocket watch of his own, which he brought along and laid on the table alongside Wheeler's watch, as if to say that his time was just as valuable as Wheeler's, even if it was measured on a cheap watch.

If Wheeler had really been the pompous ass he was pretending to be, or if Feynman had gone along with the pomposity without questioning it, their relationship might never have developed beyond the formal. As it was, both men saw the humour of the situation, and collapsed into fits of laughter reminiscent of the scenes round the dinner table at Far Rockaway – corpsing, like actors unable to continue with their lines. Every time they tried to get down to business, one of them would start giggling again and set the other off. The two men became firm friends, and when the time came Feynman had no hesitation in choosing Wheeler as his thesis adviser. The pattern of their first encounter continued throughout their student–teacher relationship: 'Discussions turned into laughter, laughter into jokes and jokes into more to-and-fro and more ideas.'[13]

A graduate student at Princeton had plenty of choice in his work, both of his thesis supervisor (if the professor he wanted was willing to take him on) and in the courses he attended. In fact, there were no formal course requirements at all (sheer bliss after the labours of English and philosophy at undergraduate level), although the student had to pass tough preliminary examinations, complete a satisfactory thesis based on original research, and defend that thesis in a rigorous oral examination. Among the classes Feynman chose to attend was a graduate-level course in biology, a subject he was to dabble in at an even higher level later in his career; there was, quite frankly, nothing he could learn from the graduate courses in physics. The research students helped each other out with their problems, though, and that way they learned a lot about what was going on in physics in general, not just the area covered by their own thesis topic. On one occasion early in his time at Princeton Feynman was able to calculate, using quantum theory, the value of a parameter that one of his fellow students needed in order to explain certain details of the way an atomic nucleus captures an electron, in the process known as inverse beta decay. It was the first time that he had made a calculation that was needed in connection with a current experiment at the cutting edge of physics.

Just as he hadn't been worried that the Greeks had discovered the rules of geometry before he had, Feynman wasn't concerned about what use, if any, his fellow student made of the calculation. 'The important thing was that I did it, that was the beginning of the real stuff, and it felt good.'[14] As ever, the important thing, to Feynman, was solving the problem. Throughout his career, he would be almost entirely unconcerned about publishing his discoveries. The important thing was that *he had done it.* He couldn't resist problem solving, and when faced with a problem he was largely unconcerned about whom he was talking to. As a research student, he had no hesitation in questioning even Albert Einstein, who by then was based at the Institute for Advanced Study, in Princeton, and gave a seminar at the university. The name and the reputation didn't mean a thing. There was just some guy, a fellow scientist, giving a talk, and if something he said didn't sound right then Feynman would question him until it did make sense.

There was another way in which Feynman lacked respect (in the best possible way) for authority, linked to his love of problem solving. He wanted to work out *everything* for himself, from first principles.[15] That way, he could be sure he had got it right, instead of, perhaps, wasting valuable time developing someone else's ideas, only to find that those ideas had been wrong in the first place. He was encouraged in this attitude by the last sentence in the 1935 edition of Dirac's book on quantum physics, which said, 'it seems that some essentially new physical ideas are here needed' – a sentence he quoted to himself as a kind of mantra for the rest of his life. Whenever Feynman was stuck with a physics problem he was working on, even in the 1980s, he would walk around muttering 'it seems that some essentially new physical ideas are here needed' while trying to find a way out of the impasse.[16]

The sentence made such an impression on Feynman, when he first read it, because Dirac himself was admitting that quantum theory as understood in the 1930s was incomplete and imperfect, and that new ideas were needed. So, surely, the last thing to do was to try to use these old ideas as the starting point for a new version of quantum physics. Better, thought Feynman, to start entirely from scratch, build his own quantum theory, and see if the problems that so puzzled Dirac and his contemporaries could be solved that way. This idea had been firmly planted in Feynman's mind while he was still at MIT; it flowered at Princeton, and came to fruition, as we shall see, in his masterwork after the Second World War.

But all that still lay far in the future when the young graduate student was struggling to come to terms with the social scene at Princeton. Princeton was deliberately designed as an imitation of the old colleges of Oxford and Cambridge, both in its architecture and its social style – and its imitation English accents. Feynman was assigned a room in the Graduate College, an impressive, ivy-clad building complete with a Great Hall with stained glass windows and a Hall Porter to guard the door of the college. He was more than a little nervous about what he had let himself in for, especially since his colleagues at MIT had been teasing him about how horrified Princeton would be by this rough diamond from Brooklyn. He had hardly got settled in his room, on the Sunday he arrived, when he was invited to join the Dean for tea – a regular Princeton ritual – that afternoon. The event was very formal, Feynman was very nervous and didn't know anybody. It was there, his mind preoccupied with trying to work out where he should go, and whether he ought to sit down, that he was offered tea by the Dean's wife, who asked if he would like cream or lemon in it. 'Both please', he replied absentmindedly, leading to the famous response 'Surely you're *joking*, Mr Feynman!' that later became the title of his first bestselling popular book.

But Princeton wasn't all formality and imitation English manners; it had a first-class physics school, to which Feynman had been attracted by noticing how often the address appeared on papers in the *Physical Review*. He imagined the Princeton cyclotron (an early form of particle accelerator) as a huge and impressive instrument, polished with care and attended by acolytes in gleaming white coats (rather like the 'scientists' you see in a soap powder commercial today). But when he went over to the physics building, the day after the Dean's tea incident, to see the great machine for himself, he found something else entirely; a homely device tucked away in a basement and surrounded by wires and cables and water pipes, with bits of wax stuck over it where things were being fixed, and water dripping from some of the pipes. It was just like his own childhood 'laboratory' on a larger scale – a real research instrument, where people tinkered 'hands on' and persuaded the machine to perform its tricks. Nothing could be more different from the formal face of Princeton typified by the Dean's tea, and Feynman fell in love with the cyclotron on the spot, happy to be reassured that he had indeed come to the right place to do his kind of physics.

Graduate College also had its advantages, since people from all disciplines were living together under one roof, and Feynman could get involved in deep discussions with researchers from other fields. Sometimes he sat at dinner with the philosophers (winding them up by demonstrating the shallowness of their debates), sometimes with the biologists and often with the mathematicians. He learned that he was able to keep track of time accurately for long periods by counting in his head, and competed with John Tukey, who later became an eminent statistician, in performing this trick while engaged in other tasks, such as reading, or running up and down stairs. They discovered that they did their mental counting in different ways. Feynman 'heard' a voice counting off the seconds in his head, while Tukey 'saw' the numbers marching past. As a result, Feynman could read a book while still keeping his mental count, but Tukey could not, because the reading part of his brain was busy. On the other hand, Tukey could talk while counting, but Feynman could not (so he could not read out loud), because the verbal part of his brain was busy.[17] It was only much later that Feynman realized that this was an important discovery about the working of the mind, showing how the same end could be achieved in different ways, and was original enough for it to have been published in a psychology journal in the 1940s.[18]

At Wheeler's house, where Feynman often worked with Wheeler, he would amuse Wheeler's two small children with jokes and tricks, including demonstrating how to tell if the contents of a tin can are liquid or solid by tossing the can in the air and watching the way it wobbles in its flight.[19] When a professor of psychology visited Princeton to lecture on hypnosis, Dick was the first to volunteer to be hypnotized (to his surprise, it worked). And his drumming became more practised.

Feynman was happy, his work was going well (as we discuss in the next chapter) and in many ways the future seemed assured. In spite of Melville's fears (he also called on Wheeler during Richard's time at Princeton, this time asking outright whether anti-Jewish prejudice might affect Richard's career and being firmly told 'no'), there would clearly be no problem about Richard's finding a job after completing his PhD, and as soon as he was no longer a student he and Arline could marry. Long before the end of Richard's first year at Princeton, the authorities were well aware that they had something special on their hands. In a reference endorsing Feynman's application for a Proctor Fellowship, on 17 May 1940, H. P. Robertson, the Professor of Mathematical Physics, described him as a 'most promising student' and said that 'at the corresponding stage in their careers', Richard's showing was 'better than that of John Bardeen'.[20] Bardeen later became the first person to win the Nobel Prize in Physics twice, which is some indication of how special Feynman was as a graduate student. He would certainly have no problem making a career in physics. But there were two clouds on the horizon.

The war in Europe had begun almost exactly at the time Feynman was joining Princeton, in the autumn of 1939. Over the following months, the United States became increasingly involved in the war effort as a 'neutral' supporter of the British cause, and it seemed increasingly likely that at some stage the country would formally join in the fight against Hitler. Few of Feynman's peers had any doubts that this would be the right thing to do; many eminent Jewish scientists (including Einstein) were now based in America precisely because they had had to flee from Hitler's Germany, and the war could be understood in black and white terms (and was, indeed, being presented that way in government propaganda), as 'good guys against bad guys'.

Like many of his contemporaries, Feynman increasingly felt that he ought to do something to help the war effort. When he had been at MIT, he had repeatedly tried to get a summer job at the Bell Laboratories, failing (it seems likely with hindsight) because of an unspoken objection to his Jewish background. In the spring of 1941, at the fourth or fifth time of asking, he was accepted, and felt very happy. But then a General visited Princeton, to give a talk about the importance of physics in the modern army, exhorting the young physicists to take up war work. Carried away by the patriotic fervour, Feynman gave up the opportunity to work at Bell Labs (even though they offered to give him war-related work), and spent the summer instead at the Frankfort Arsenal, in Philadelphia, working on a mechanical detector (a kind of primitive computer) to be used by artillery specialists to predict where an aeroplane would be by the time the shells fired from the guns reached its altitude. He was so successful that the army offered him a long-term job at the end of the summer, as head of his own design team. But in September he went back to Princeton to finish his PhD. He hadn't enjoyed the army's bureaucratic way of doing things, and felt that he should have gone to Bell Labs after all. But the experience of military research in the summer of 1941 was soon to stand him in good stead.

Finishing the PhD was, of course, simply a formality. Feynman had completed the preliminary requirement, the qualifying examination, a year earlier, in the autumn of 1940, having spent most of the summer at MIT (where he could work without disturbance, in the library) studying in preparation for the event. He had indeed carried out original research work, with John Wheeler, and all that was left was to write up his thesis satisfactorily (even Feynman realized that he would be able to defend it adequately). But the outside world intervened. The Japanese attack on Pearl Harbor brought the United States into the war, against both Japan and Germany, on 8 December 1941.

One morning in December, one of the Princeton physicists, Robert Wilson, came into Feynman's office. He told Dick that he had a secret to divulge, something that Wilson wasn't supposed to tell Dick, but which he would anyway, because he knew that once Dick heard the story he would join Wilson's Top Secret project. He explained about the possibility of building an atomic bomb, the fear that Germany might already be working on such a project, and the need to find a technique for separating the radioactive uranium-235 that would be needed for such a bomb from the more common, stable variety, uranium-238. Wilson's project involved a way of carrying out this separation (not the technique that was actually used, in the end, to make the first bombs), and he wanted Feynman to join the team.[21]

Feynman's first reaction was to say no. He had had enough of military bureaucracy. He would keep the secret, but he had to finish his thesis before he could think of doing anything else. Wilson got up and quietly left the office, telling Dick that if he changed his mind, he would be welcome to attend a meeting in Wilson's office at 3 o'clock that afternoon. Feynman tried to get back to work on his thesis, but couldn't. He thought about the possibility of Hitler obtaining a nuclear bomb, and the implications. He remembered the stories he had heard from the refugees from Hitler's Germany. At 3 o'clock, he was at the meeting in Wilson's office. By 4 o'clock, he had his own desk in another office and was at work, the thesis temporarily abandoned in the drawer of the old desk. It was the right decision. All science in the United States stopped during the war, except for what became the Manhattan Project – 'and that', said Feynman in *Surely You're Joking*, 'was not much science; it was mostly engineering'.

Feynman worked for a few months on the uranium separation project, but it wasn't moving as fast as had been hoped, so in the spring of 1942, urged on by Wheeler (who was by now working with Enrico Fermi in Chicago on the design and construction of the world's first artificial nuclear reactor, and warned Dick that this might be his last chance to finish the PhD before becoming totally embroiled in war work), he took a few weeks' leave to finish his thesis. Then, after submitting the thesis, he went back to work with Wilson's team. The thesis examiners, John Wheeler and Eugene Wigner, described it as 'exceptionally original', and the official report of his oral examination, held on 3 June 1942, rated his performance as 'excellent'.[22] Feynman formally received his PhD degree at the regular Princeton commencement ceremony in June 1942, attended by his proud parents. But one important person was not present. Arline was ill in hospital with what had recently (but belatedly) been diagnosed as tuberculosis. According to Wheeler, her tragic illness may have been exacerbated by her overtaxed lifestyle, burning the candle at both ends, being a full-time art student in New York by day, teaching piano in the evenings to pay for her course, and visiting Richard at Princeton whenever possible for weekend dances.[23] In spite of the serious nature of her illness, the couple married before the end of the month, fulfilling their long-standing promise to each other.

The relationship between Richard and Arline had initially continued while he was at Princeton in very much the same way as it had gone on while he was at MIT. They spent a lot of time together during the vacations, and she visited him increasingly often at

Princeton. It was Arline who, around this time, would try to shake Richard out of any gesture he made towards conformity by reminding him of what he had said to her many times before: 'What do *you* care what other people think?' The slogan became the title of his second bestselling book in the 1980s. But just when the future looked assured for the couple, Arline developed a lump in her neck. It wasn't painful, but she began to feel increasingly tired – Joan Feynman recalls[24] an occasion when the Feynman family went to stay in Atlantic City for a holiday, and Dick joined them for the weekend. Everyone was swimming in the pool, having a great time, but Arline had to get out and lie down because of her tiredness.

After a while, the lump changed slightly, Arline developed a fever and was taken into hospital. The illness was initially misdiagnosed, first as typhoid, then as Hodgkin's disease. Meanwhile Feynman, reading up the symptoms in the medical library at Princeton, decided that they matched those of tuberculosis of the lymphatic glands. But the textbook said 'this is very easy to diagnose', so he decided that that couldn't be the problem, or the doctors would have found it.[25] Either way, Hodgkin's disease or TB, the disease was, in those days, almost certainly fatal. Feynman wanted to tell Arline the truth, but was prevailed upon by her family to join in a 'white lie' that it was only glandular fever, and that she would soon recover.

For a time, Arline did get slightly better, and returned home, where she guessed that the illness was much more serious when she heard her mother crying. To his immense relief, Richard was able to tell her the truth (as the doctors saw it at the time), and the couple began to re-plan their future in the light of this new development.

By now, Richard was nearing his final year at Princeton, and had been awarded a scholarship, one of the stipulations of which was that it could not be held by a married student. His first reaction to the news of Arline's death sentence – she had been given a maximum of two years to live – was that he wanted to marry her at once, and look after her for her final years. Incredibly, when Feynman asked permission for the 'no marriage' rule to be waived in this very special case, it was refused. He would have to choose between the scholarship and marriage, and without the scholarship he couldn't afford to live while completing his PhD. Feynman seriously considered giving up the thesis and seeking a job with Bell Labs, or somewhere similar. But then, at last, came the final, correct diagnosis of Arline's condition. It was indeed tuberculosis of the lymphatic glands.

The situation by then was so desperate that, in spite of Feynman's anger at his own failure to press the possibility of this diagnosis on the doctors (although it is hard to see how they would have taken much notice of him), the couple regarded this diagnosis as good news. After all, according to the doctors it might mean that Arline would live for another five years. So the immediate pressure to marry was eased, and Richard did keep the scholarship, finish his PhD and get started on the road that would lead him to Los Alamos to work on the Manhattan Project.

Feynman came under enormous pressure from his family and friends not to go through with the marriage. Chief among the opposition, Melville still thought that marriage even to a healthy bride would damage Richard's career prospects, while Lucille was more concerned that he would catch TB and die. But there was never any doubt in Richard's own mind that he was doing the right thing, even though he and Arline understood that they could only ever have a limited physical relationship, and could not even kiss for fear of contagion. The decision to go ahead with the marriage as soon as he had received his PhD led to a rift between Richard and his parents which was never really healed, but he never regretted the decision.

By the time Dick had been awarded his PhD, Arline was permanently hospitalized, staying at the state hospital on Long Island. He arranged for her to move to a charitable hospital called the Deborah Hospital, at Browns Mills (near Fort Dix), in New Jersey, close to Princeton. He borrowed a stationwagon from a friend at Princeton, and fixed it up, as he described in *What Do You Care*, 'like a little ambulance', with a mattress and sheets in the back for Arline to rest on. On 29 June 1942, the couple had a romantic ride on the Staten Island Ferry, and got married in the borough of Richmond (sealing the ceremony with a chaste kiss on the cheek); then the groom delivered his bride to the Deborah Hospital, and left her there, where he could visit her every weekend.

For a while, the newly married, newly qualified Dr Feynman stayed on at Princeton, working on the fringes of what was now called the Manhattan Project. Arline, an indomitable character, kept her spirits up in hospital by writing daily letters to Dick, initiating crazy projects (more of these in Chapter 5), and planning for what she knew was a mythical future of normal married life. Once, she sent Dick a box of pencils, each one emblazoned in gold with the message 'RICHARD DARLING, I LOVE YOU! PUTSY'. He was pleased, but embarrassed. The message was nice, and (as Arline had known) he needed pencils. Such stuff was in short supply, and too valuable to waste. But he didn't want one of the professors noticing the legend. So he got a razor blade and neatly cut the message off of the pencil he was using. But Arline was ahead of him.

Next morning, he received a letter from her, beginning 'WHAT'S THE IDEA OF TRYING TO CUT THE NAME OFF THE PENCILS?', and ending 'WHAT DO *YOU* CARE WHAT OTHER PEOPLE THINK?' He left the rest of the pencils intact, ignoring the gentle ribbing that resulted when colleagues picked them up. It was a message that he took to heart, carrying it with him to Los Alamos and beyond, right up to his involvement with the *Challenger* inquiry. But before we go into all that, it is time to take stock of the scientific work with which Feynman made his initial reputation, as an undergraduate at MIT and then as a research student at Princeton.

Notes

1. Mehra.
2. Mehra.
3. Quoted by James Gleick in *Genius* (see Bibliography).
4. From an unpublished memoir by Welton, written in 1983; quoted by Mehra.
5. Mehra.
6. Mehra.
7. *Surely You're Joking.*
8. Mehra.
9. *Most of the Good Stuff.*

10. See letters between Smyth and Morse, and Smyth and Slater, Princeton archive (Seeley G. Mudd Manuscript Library).
11. Philip Morse, *In at the Beginnings*, MIT Press (1977).
12. Joan Feynman, correspondence with JG, January/February 1996.
13. Wheeler, in *Most of the Good Stuff.*
14. Mehra.
15. This was, in fact, a characteristic he shared with Einstein; see *Einstein: A Life in Science.*
16. Ralph Leighton, comment to JG, December 1995.
17. *What Do You Care.*
18. Leighton, comment to JG, December 1995.
19. Wheeler, in *Most of the Good Stuff.*
20. Princeton archive.
21. Mehra.
22. Report in the Princeton archive.
23. See Wheeler's contribution to *Most of the Good Stuff.*
24. See *No Ordinary Genius.*
25. *What Do You Care.*

— 4 —

Early works

Although Feynman failed to get much intellectual stimulation from being forced to attend classes in English and philosophy during his time as an undergraduate at MIT, he had ample opportunity to stretch his mind by attending any classes that did interest him, even if they did not count officially towards his degree. One of the classes he attended in his senior year was taught by Manuel Vallarta, who had an interest in cosmic rays – high-energy particles that reach the Earth from space. These 'rays' come equally from all directions – they are isotropic – but the stars of our Galaxy, the Milky Way, are distributed far from uniformly across the sky. The obvious inference is that cosmic rays do not come from within our Milky Way Galaxy, but from the Universe at large, beyond the Milky Way. But even if cosmic rays did come in to the Galaxy uniformly from all directions, surely, Vallarta thought, they ought to be scattered by the stars of the Milky Way, and end up with an uneven pattern on the sky. He discussed the puzzle with Feynman, and suggested that the bright undergraduate might like to work on the puzzle of the isotropy of cosmic rays.

Feynman was able to solve the puzzle in a fairly straightforward manner, proving that if cosmic rays from the Universe at large do indeed enter our Galaxy isotropically, then they will still be seen coming from all directions when they reach the Earth. The influence of the stars of the Milky Way is far too small to disturb the pattern. One interesting feature of Feynman's proof is that it involved dealing mathematically not just with cosmic ray particles coming into the Galaxy from outside, but with a kind of hypothetical mirror image set of particles moving out from the Galaxy into deeper space. The kind of scattering Vallarta was worried about mainly involves the magnetic fields of stars, which interact with electrically charged particles. So the probability of an electron (with negative charge) coming into the Galaxy along a particular path is the same as the probability of a positron (with positive charge) going out of the Galaxy along the same path.

In fact, some cosmic rays are now known to originate within our Galaxy, but that does not affect the validity of Feynman's argument – those cosmic rays which do originate from outside the Milky Way (essentially the ones with most energy, which is why they were the first to be studied) behave just as he calculated on their way to Earth. Vallarta was sufficiently impressed by Feynman's proof that he offered to tidy it up and submit it to the *Physical Review* for publication, under their joint names. He explained to Feynman that although Vallarta had made only a small contribution to the paper, his name should appear first on it, because he was the more senior scientist. It was Feynman's first experience of this kind of jockeying for academic credit, but he was hardly in a position to object, and the paper appeared in the *Physical Review* on 1 March 1939, with the authorship 'Vallarta and Feynman'.

But Feynman would have the last laugh. In 1946, Werner Heisenberg published a book on cosmic rays in which he discussed just about every worthwhile paper ever published on the topic. The Vallarta and Feynman paper didn't quite fit in anywhere, but right at the end of the book Heisenberg discussed the possibility of the influence of stellar magnetic fields in changing the direction of cosmic rays and, in his very last sentence, concluded that 'such an effect is not expected according to Vallarta and Feynman'. The next time Feynman met up with Vallarta, he gleefully asked if he had seen Heisenberg's book. Vallarta already knew what was coming. 'Yes', he said. 'You're the last word in cosmic rays.'[1]

Feynman had time to do research – albeit in a modest way – in his senior year because by then he was only serving out time as far as the requirements for his degree were concerned. The rules said you had to serve four years as an undergraduate before receiving the Bachelor's degree. Feynman had long since learned everything required of a physics student, and more, but he hadn't completed the statutory four years. In fact, unknown to Feynman at the time, Philip Morse had actually suggested to the authorities at MIT that

Feynman should be allowed to graduate a year early, after three years instead of four; but the proposal had been turned down. All that was left was for Feynman to write his senior thesis – no small task at the end of the 1930s, when students were expected to do original work on a specific problem suggested to them by a supervisor. The supervisor was supposed to be aware of the broad sweep of the development of science, and able to pinpoint a tiny area, equivalent to adding one brick to the tower of knowledge, where the undergraduate could make a genuine contribution. Feynman's senior thesis started out like that, but ended up as a much more far-reaching piece of work.

The problem John Slater set Feynman was to work out why quartz expands much less than other substances, such as metals, when it is heated. Feynman quickly became intrigued by the whole idea of how and why things expand at all, and set out to study the way the forces between atoms work in crystals.

In a crystal, the atoms are spaced out at regular intervals in a three-dimensional array, or lattice. They are held in place by electrical forces, but tend to jiggle about a bit. When a crystal is heated, there is more jiggling and the spacing between atoms increases slightly, which is why the crystal expands.

Once he started thinking about how the forces between atoms alter as the crystal expands, Feynman also became intrigued by the behaviour of those forces when the crystal is compressed, so that the atoms are squeezed closer together. He realized that he could treat the forces between pairs of atoms (not just in crystals but also in molecules) as acting like little springs. A spring resists being stretched, but it also resists being compressed. Some of this work covered ground that had already been covered by other people, but Feynman didn't know that, and worked everything out for himself, from first principles, in his usual way. The basis of Feynman's approach was that the force on any nucleus in a molecule or in a crystal lattice can be worked out from the distribution of electrical charge on nearby nuclei and in the electron clouds surrounding the nuclei simply from classical electrostatics, once the distribution of the electron cloud is known. You still need quantum mechanics to work out the distribution of electrical charge in the cloud, but once you have done that the rest is, relatively speaking, plain sailing.

After some elegant and sophisticated manipulation of the relevant equations, Feynman was able to prove that the force on each nucleus could be calculated from a relatively simple expression, saving an enormous amount of labour in carrying through these kinds of calculations. His senior thesis, entitled 'Forces and Stresses in Molecules' and running to just 30 double-spaced pages of typescript, impressed Slater sufficiently for him to encourage Feynman to write it up in a slightly different form for the *Physical Review*, where it appeared under the title 'Forces in Molecules' later that year (1939). The simplification which so greatly eased the burden of work for chemists trying to calculate the behaviour of atoms in molecules and crystals was also discovered, independently, by another researcher, and is known as the Feynman–Hellman theorem. It is still used today – not bad for a piece of undergraduate thesis work dating back more than half a century.

One of the most striking things about the senior thesis, though, is that the elegant manipulation of equations is set in a clear, no-nonsense and jargon-free text that carries the authentic Feynman 'voice', reading almost like a transcript of a talk. He clearly knew not only how to do physics, but also how to explain physics, at an early age.

By the time Feynman arrived at Princeton, he was more than ready for full-time work in research, and he found in Wheeler exactly the right kind of supervisor to stimulate him in the development of much more original ideas about the way the world works. Early in their relationship, Wheeler set Feynman a few fairly straightforward problems to investigate. His student's success at these helped to establish, as if Wheeler had not already realized it, that he was dealing with a special talent. At the same time, Feynman learned how much Wheeler already knew about quantum mechanics. All the while, Feynman was also puzzling over an idea that he had been working on, intermittently, as an undergraduate at MIT. Soon, he was ready to air the puzzle with his new mentor.

Feynman's jumping-off point, as he stressed when he gave his Nobel lecture in Stockholm in 1965,[2] was the conclusion of Dirac's 1935 book – 'it seems that some essentially new physical ideas are here needed'. Nowhere was this need for new ideas more obvious than in the puzzle of what was called the 'self-energy' of the electron.

It all had to do, as far as the undergraduate Feynman could tell, with the concept of a field of force. A charged particle, such as an electron, was supposed to interact with other charged particles by being surrounded by a field of force. The field gets weaker the further away you are from an electron, so it interacts most strongly with nearby charged particles. But, embarrassingly, there is no limit to how strongly the field can interact, provided you get close enough to its source. In fact, the strength of the interaction is inversely proportional to the square of the distance involved. But an electron is a point charge – it has zero radius. So at the electron itself the strength of the field would be 1 divided by zero, which is infinity. In other words, each electron should have an infinite self-energy – which, among other things, would give it infinite mass, in line with Einstein's equation $E = mc^2$.

This version of the problem arises even without invoking quantum mechanics; the problem also arises, even more forcefully, in the context of quantum theory. As an undergraduate, Feynman surmised that the answer to the problem must be that the electron does not act on itself at all; from there, it was a small step for him to reject the whole notion of a field, even though the field concept lay at the heart of physics. The prevailing field theories said that if all the charges combine to make a single common field, and the common field interacts with all the charges, there is no way to avoid each charge interacting with itself.

Feynman's idea was to go back to the older concept of action at a distance, a direct interaction between charges, albeit with a delay.* On this picture, one electron shakes, and a certain time later another electron shakes as a result (the time delay depends on the distance to the second electron and the speed of light). But there is no way for the first electron to interact with itself. This was the state of the idea when Feynman arrived in Princeton. He hadn't worked out a proper theory along these lines; it was no more than a half-baked idea. But, as Feynman recounted in Stockholm in 1965, he had fallen 'deeply in love' with the notion, and 'I was held to this

*Unfortunately, the word 'action' is used in two different ways by physicists. This 'action at a distance' has nothing to do with the 'action' that appears in the Principle of Least Action, but is shorthand for 'interaction at a distance'.

theory, in spite of all difficulties, by my youthful enthusiasm' ('youthful enthusiasm', of course, sums up Feynman's approach to all of his work, and life in general, whatever his chronological age). There was, though, a 'glaringly obvious' fault with the idea, as he pointed out in that Nobel address. A charged particle such as an electron *must*, in fact, interact with itself to a certain extent, to account for a phenomenon known as radiation resistance.

All objects resist being pushed about – this is the property known as inertia. In a frictionless environment, such as the inside of a spaceship falling freely in orbit around the Earth, any object will sit still (relative to the walls of the spaceship) until it is given a push, then it will keep moving at a steady speed in a straight line (that is, at constant velocity) until it is given another push (perhaps by bouncing off the wall). The key thing is that it takes a force to make something accelerate – which, to a physicist, means to change its speed or its direction of motion, or both. This is encapsulated in Newton's Laws of Motion, which became the basis of classical mechanics more than 300 years ago, and which still provide an entirely adequate description of the way things work for most everyday purposes, whether that involves designing a bridge that won't fall down or a spaceship that will fly to the Moon.

Just why things have inertia – where inertia 'comes from' – is not explained by Newton's laws, and Einstein tried to build inertia into his General Theory of Relativity, without entirely succeeding (but see Chapter 14). But that doesn't matter for now. What does matter is that if you try to accelerate a charged particle, perhaps by shaking it to and fro with a magnetic field, you discover that it has an extra inertia, over and above the inertia you would find for a particle with the same mass but no electric charge. This extra inertia makes it harder to move the charged particle.

Now this is not just some exotic phenomenon only of interest to physicists. The most common reason for shaking electrons to and fro is to make them radiate electromagnetic energy, in line with Maxwell's equations. This is what goes on in the broadcast antennas of TV and radio stations. It takes energy to make the electrons in the antenna oscillate and radiate the signal you want to broadcast, and it takes more energy (requiring a more powerful transmitter) than it would to shake equivalent uncharged particles, which do not radiate, by the same amount. Hence the name radiation resistance. The effect of radiation resistance can be seen in the electricity bills of every TV and radio station.

One curious feature of the classical description of electrons (the same is true for all other charged particles) and electromagnetic fields is that the interaction between each electron and the field (the self-interaction) actually has two components. The first component looks as if it ought to represent ordinary inertia, but is infinite for a point charge. But the second term exactly gives the force of radiation resistance. So the snag with Feynman's original idea, that an electron could not act on itself at all, was that even if the idea could be made to work it would remove both terms in the expression, getting rid not just of the unwanted infinity but also of the radiation resistance. This was the state of play when he started thinking seriously about the idea once again at Princeton.

Feynman needed some interaction to act back on the electron and give it radiation resistance when it was accelerated, and he wondered whether this back-reaction might come from other electrons (strictly speaking, any other charged particles) rather than from the 'field'. As physicists do when trying to get to grips with such problems, he considered the simplest possible example – in this case, a universe in which there were only two electrons. When the first charge shakes, it produces an effect on the second charge, which shakes in response (this, of course, is how the receiver in your radio or TV set works, as electrons in it respond to the shaking of the electrons in the broadcast antenna). But now, because the second charge is shaking there must be a back-reaction which shakes up the first charge. Perhaps this could account for radiation resistance. Feynman calculated the size of the effect, but it didn't work out properly to account for radiation resistance. Baffled, but still in love with the idea, it was at this point that he took it to Wheeler to discuss.

What Feynman didn't know was that Wheeler had been interested in the idea of action at a distance for some time, and that this had a respectable pedigree as a backwater of physics.[3] So the professor didn't dismiss his student's idea as crazy, but set out with him to work through the calculations. To Feynman's embarrassment, Wheeler pointed out the big flaw with his calculation. It takes a certain time for the second electron to respond to the shaking of the first electron, and the same amount of time before the first electron responds to the shaking of the second electron. So the reaction back on the first electron would occur some time after it had been shaken in the first place – not at the right time to cause radiation resistance. What Feynman had actually described and calculated, albeit in an unconventional manner, was simply ordinary reflection of light.

But Wheeler didn't stop there. Maxwell's equations, he pointed out, actually have two sets of solutions. One corresponds to a wave moving outward from its source and forward in time at the speed of light; the other (usually ignored) corresponds to a wave converging on its 'source' and moving backwards in time at the speed of light (or, if you like, moving forwards at *minus* the speed of light, $-c$). This is rather like the way in which the equations of quantum mechanics can be solved to give a solution corresponding to positive-energy electrons and a solution corresponding to negative-energy electrons. Dirac's equation, published in 1928, was still the crowning glory of quantum mechanics at the beginning of the 1940s, so it did not seem completely crazy for the two young physicists to take the second solution to Maxwell's equations seriously. The waves corresponding to the usual solution of the equations are called retarded waves, because they arrive somewhere at a later time than they set out on their journey (the journey time is 'retarded' by the speed of light); the other solution corresponds to so-called advanced waves, which arrive before they set out on their journey (the journey time is 'advanced' by the speed of light). If the back-reaction from the second electron only involved advanced waves, Wheeler realized, its influence on the first electron would arrive exactly at the right time to cause radiation resistance, because it would have travelled the same distance at the same speed, but backwards in time.

Wheeler set Feynman the task of calculating what mixture of advanced and retarded waves would be required to produce the correct form of radiation resistance. Between them, Wheeler and Feynman also proved that in the real Universe, full of charged particles, all the interactions would cancel out in the right way to produce the same radiation resistance that they had calculated for the simple case.

A key ingredient of their model is that a wave has both a magnitude and a 'phase' – if two waves are the same size, but one is precisely out of step with the other, so that the first wave produces a peak where the second wave produces a trough, they are out of phase, and cancel each other. If the two waves march precisely in step, so that the two peaks are on top of each other, they are in phase, and produce a combined wave twice as big as either wave on its own. As a result, it turned out that you need a mixture of exactly half advanced waves and half retarded waves generated by each charge every time it shakes (see Figure 5), using the solution of Maxwell's equations that is completely symmetrical in time.

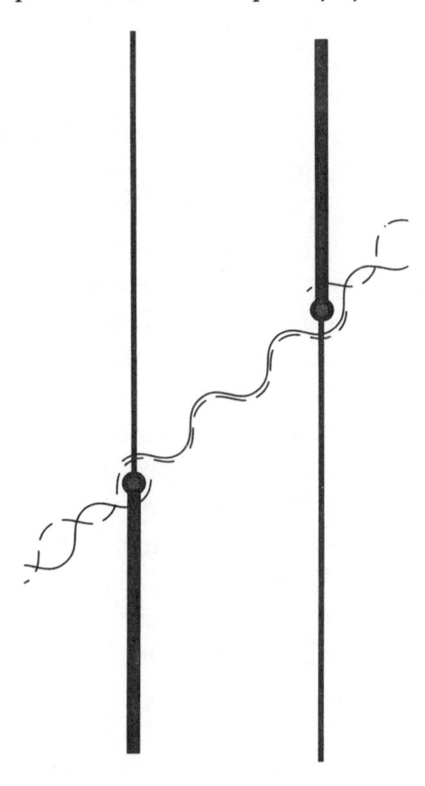

Figure 5. The Wheeler–Feynman theory of radiation describes the interaction between two charged particles in terms of waves moving forwards and backwards through time. Because of a phase change at the charged particles, the waves exactly cancel out everywhere except in the region of spacetime between the particles, where they reinforce one another. See also Figure 3, p. 30.

Wheeler discovered that the Dutch physicist Adriaan Fokker had reached a similar conclusion in a series of papers published between 1929 and 1932; but Feynman's version was much more straightforward and easier to understand, while Fokker had never developed his ideas further. The half wave which is retarded goes out from the first electron forwards in time, while the half wave that is advanced goes out backwards in time. When the second electron shakes in response, it produces another half retarded wave which is exactly out of step with the first wave, and so precisely cancels out the remaining half retarded wave for all later times, and a half advanced wave which goes back down the track of the first wave to the original electron, in step with that wave, reinforcing the original half wave to make a full wave matching the usual solution to Maxwell's equations. This half advanced wave arrives at the first electron, of course, at the moment it starts to shake, and causes the radiation resistance. Then it continues back into the past, cancelling out the original half advanced wave from the first electron. The result is that between the two electrons there is a single wave exactly matching the conventional solution to Maxwell's equations, but everywhere else the wave cancels out, and radiation resistance emerges automatically from the equations, while the infinite self-energy never appears.

'If we assume all actions are via half-advanced and half-retarded solutions of Maxwell's equations and assume that all sources are surrounded by material absorbing all the light that is emitted, then we could account for radiation resistance as a direct action of the charges of the absorber acting back by advanced waves on the source.'[4] Because of this essential role of the absorber in determining the way radiation is emitted, it is sometimes called the 'absorber theory' of radiation.

It took months to work all this out, in the autumn of 1940. In its initial form, as we have described it here, the theory still involved electromagnetic waves, an echo of the field that Feynman had been trying to do away with; but the two researchers also found, to Feynman's delight, that the whole thing could indeed be described without using Maxwell's equations at all, but directly in terms of the motion of the particles involved and a suitable time delay, using the Principle of Least Action, without any vestige of a field at all. This only works if the interaction is half advanced and half retarded, when it turns out that interactions can only occur after delays corresponding to influences which travel at the speed of light. All of conventional electrodynamics could be written in this new and mathematically simple way, without involving electromagnetic waves or fields at all, provided you were open-minded enough to accept the reality of interactions that travelled backwards in time – that when one electron shakes, another electron may shake as a result *before* the first electron shakes. And, as Feynman pointed out in his Nobel lecture, aside from gravity, electrodynamics 'is essentially all of classical physics'. It was another example of the way in which fundamental features of physics could be described in quite different ways to give the same answers.

While the two researchers were working on all this in the autumn of 1940, one day Feynman received a telephone call from Wheeler. He said, 'Feynman, I know why all electrons have the same charge and the same mass.' When Feynman asked, 'Why?', he replied, 'Because they are all the same electron!' And he explained his latest bright idea, that a positron could be regarded as an electron going backwards in time, and that all the electrons and all the positrons in the Universe were really a kind of cross-section through a complicated zig-zag path in which a single particle traversed the Universe, through both space and time, in a complicated knot. When the first flush of his enthusiasm had worn off, Wheeler found that the idea couldn't really be made to work, not least because there would have to be the same number of positrons in the Universe as there are electrons, since for every zig forwards in time there must be a corresponding zag backwards in time. In fact, there don't seem to be any positrons in the Universe except ones created in particle interactions, that soon meet up with electrons and annihilate. But Wheeler's bright idea contained the germ of an important concept which Feynman would later develop in a different way – the idea that changing the direction in which an electron is moving through time is equivalent to changing the sign of its charge, so that an electron going forwards in time *is* a positron going backwards in time, and vice versa. Positrons could simply be represented, in all quantum mechanical calculations, as electrons going from the future to the past, like the advanced waves in the usually neglected solution to Maxwell's equations: yet another example of how the same thing could be described in different ways.[5]

Wheeler decided that Feynman's next task, in the spring of 1941, should be to give a talk describing the work on direct action at a distance and time-symmetric electrodynamics. Learning how to present work in front of your peers in seminars is an essential part of the training of a research student. Although the first talk is always a fairly informal occasion, for the benefit of people at the student's home institution, it is a nerve-racking occasion for most students. Feynman not only had to present a highly controversial new idea, but at Princeton in those days, even for an internal seminar, his audience would include Eugene Wigner, one of the leading quantum theorists, Henry Norris Russell, one of the greatest astronomers of the time, John von Neumann, regarded as the smartest mathematician of his generation, Wolfgang Pauli, one of the quantum pioneers who just happened to be at Princeton on a visit from Switzerland, and Albert Einstein, who worked at the nearby Institute for Advanced Study.

Feynman has described his nervous preparations for the talk in *Surely You're Joking*, and how at the start of the talk his hands shook as he pulled his notes out from the brown envelope in which he had put them for safekeeping:

> But then a miracle occurred, as it has occurred again and again in my life, and it's very lucky for me: the moment I start to think about the physics, and have to concentrate on what I'm explaining, nothing else occupies my mind – I'm completely immune to being nervous. So after I started to go, I just didn't know who was in the room. I was only explaining this idea, that's all.[6]

After the presentation, Pauli spoke up, saying that he didn't think the theory could possibly be right, and turned to Einstein to ask if he agreed. 'No', replied Einstein, softly. 'I find only that it would be very difficult to make a corresponding theory for gravitational interaction.' But he didn't think that was any reason to reject the Wheeler–Feynman theory, which was a possible way forward.

It was hardly a ringing endorsement, but the theory had stood up to its first test, and Einstein had not dismissed it out of hand. Feynman's next task was to try to find a way to develop a quantum mechanical version of the theory. At first, he hesitated about getting to grips with this, because Wheeler kept claiming that he was making strides in that direction himself, but Wheeler's efforts always seemed to take him up a blind alley, leaving the field clear for Feynman. It was his effort to develop a version of quantum theory that did away with fields and simply involved action at a distance that became the topic of his PhD thesis; partly because of his concentration on the thesis in the months that followed, and partly because of the interruption caused by war work, the absorber theory of radiation was only formally published in 1945, in the journal *Reviews of Modern Physics*, in a paper under the joint names of Wheeler and Feynman, but actually written by Wheeler in a style which Feynman thought unnecessarily complicated.[7]

Until Feynman came on the scene, the way quantum mechanics had been developed was using the Hamiltonian method, which we described in Chapter 3. This involves a wave function which describes the behaviour of quantum entities, such as electrons and photons, and differential equations which describe how the wave function changes from one instant to the next. The approach is conceptually similar to using the equations of motion based on Newton's laws to describe the way in which the position of a ball thrown through an upper-storey window changes from one instant to the next as it moves along its path. In classical mechanics, as Feynman had learned back in high school from Abram Bader, you could use the Principle of Least Action to determine the entire path of the ball, from your hand through the window, without calculating how its velocity and other properties changed at each point along the path. This, essentially, is the Lagrangian approach which Feynman had scorned as an undergraduate, perhaps because it seemed too easy for the kind of problems he was working with then. Thinking in terms of particles rather than waves, the key properties involved are the positions and the velocities (strictly speaking, their momenta, but the difference doesn't matter here) of the particles.

If you are only interested in the state of a system at a particular moment in time, it is relatively easy to set up a description of the action in terms of a function (a mathematical expression) called the Lagrangian, which depends on the velocities and positions of all the particles at that time. Starting with the Lagrangian, it is straightforward, for those who feel the need, to convert into the Hamiltonian formulation and work out the quantum mechanics in the way that people had already become used to by the early 1940s. But the action involving advanced and retarded interactions (or even retarded interactions alone) brings in the key variables at two different times, simply because when one electron shakes there is a delay before the second electron shakes. It was not at all obvious to Feynman (or anyone else) how to formulate the quantum mechanical version of the appropriate Lagrangian involving two different times.

While Feynman was struggling with this problem in the spring of 1941, one evening he went to a beer party at the Nassau Tavern in Princeton. There, as he recounted in the Nobel lecture, he got talking with a physicist who had recently arrived from Europe, Herbert Jehle. Jehle asked what Feynman was working on, and Feynman told him, ending up by asking, 'Do you know any way of doing quantum mechanics, starting with action – where the action integral comes into quantum mechanics?' 'No', Jehle replied. But he did know of an obscure paper by Dirac, published eight years previously, which made some use of the Lagrangian. He offered to show it to Feynman the next day.

Next day the two physicists went to the Princeton University library, dug out the relevant bound volume of the *Physikalische Zeitschrift der Sowjetunion* (hardly the place Feynman would have looked without being pointed in the right direction!) and went through Dirac's paper together. It was exactly the kind of thing Feynman was looking for. Under the title 'The Lagrangian in Quantum Mechanics', Dirac pointed out that quantum mechanics had been developed by analogy with the Hamiltonian approach to classical mechanics, went on to say that the Lagrangian approach seemed to be more fundamental, and pointed out the desirability of finding the counterpart in quantum mechanics to the Lagrangian in classical mechanics – just what Feynman was trying to do. What Dirac then described in that paper was a way of carrying the wave function description of a quantum system forwards in time by a tiny step, an infinitesimal amount. This doesn't sound much in itself, but physicists are used to dealing with infinitesimals, which appear in differential equations which can then be integrated up to deal with much larger (macroscopic) steps in time or space. Dirac hadn't done that; he had only found a way of taking the wave function forwards in time by infinitesimal steps. But what caught Feynman's eye was the way Dirac repeatedly said in the paper that the function he was using was 'analogous' to the Lagrangian in classical mechanics. It was an imprecise use of language that obscured, rather than clarified, what Dirac was driving at.

'What does he mean?', Feynman asked Jehle. 'What does that mean, *analogous*? What is the use of [a word like] that?' Jehle didn't know. 'You Americans!', he laughed, 'you always want to find a use for everything!' Feynman decided that perhaps Dirac meant that the two expressions were equivalent to one another, but Jehle disagreed. To find out, Feynman tried setting the expressions equal to one another and working through the simplest version of the resulting equations from Dirac's paper. It didn't quite work; he had to put a constant in as well, making the two expressions proportional to one another, not exactly equal. When he did so, though, everything fell into place, and at the end of the calculation he came out with the familiar Schrödinger equation of quantum mechanics. So he turned from the blackboard and said, 'Well, you see Professor Dirac meant that they were proportional.' To a physicist such as Feynman, the terms 'equal' and 'proportional' are more or less interchangeable, because if two mathematical expressions are proportional to one another, one expression is simply the other expression multiplied by a constant, and the effect of a constant on the equations is so trivial that it can largely be ignored.

> Professor Jehle's eyes were bugging out – he had taken out a little notebook and was rapidly copying it down from the blackboard, and said, 'No, no, this is an important discovery. You Americans are always trying to find out how something can be used. That's a good way to discover things!' So I thought I was finding out what Dirac meant, but, as a matter of fact, I had made the discovery that what Dirac thought was analogous was, in fact, equal. I had then, at last, the connection between the Lagrangian and quantum physics, but still with wave functions and infinitesimal times.[8]

In 1946, Feynman had a chance to find out what Dirac had really meant by that word 'analogous'. Both Dirac and Feynman were present at the Princeton bicentennial celebration in the autumn of

1946, and Feynman took the opportunity of mentioning the 1933 paper. Did Dirac remember the paper? 'Yes', he replied. Feynman described the functions involved. 'Did you know that they are not just analogous, they are equal, or rather proportional.' Dirac replied, 'Are they?' Feynman said, 'Yes.' Dirac said, 'Oh, that's interesting.' He really had not known, until Feynman told him, that the quantity he had described in the 1933 paper was indeed the Lagrangian required as the basis for a new understanding of quantum physics.[9]

It was only a couple of days after going through Dirac's old paper with Jehle that Feynman had the flash of insight which allowed him to use this Lagrangian to solve problems involving paths through space and time joining events a finite distance apart, instead of only an infinitesimal distance apart. These four-dimensional trajectories are known as world lines, and can be represented in two-dimensional graphs by imagining all of the three dimensions of space compressed into a single direction 'across the page' with the passage of time denoted by separation 'up the page' (see Figure 6). A line on such a diagram represents the history of a particle as it takes a certain amount of time to move from A to B, and beyond. The insight Feynman had, while lying in bed one night, unable to sleep,[10] was that you had to consider every possible way in which a particle could go from A to B – every possible 'history'. The interaction between A and B is conceived as involving a sum made up of contributions from all of the possible paths that connect the two events.

For obvious reasons, this became known as the 'sum over histories', or 'path integral' approach to quantum mechanics. A useful, if slightly imprecise, way to think of this is that the least action idea in effect gives you the integral (or sum) along a single trajectory, while the path integral approach extends this to include all possible trajectories, summing up (integrating) the paths themselves together, not just integrating along one path. We will explain the technique more fully in Chapter 6, in the context of Feynman's later work. As Feynman put it in his Nobel lecture, 'the connection between the wave function of one instant and the wave function of another instant a finite time later could be obtained by an infinite number of integrals'; and this kind of infinity is no problem to the

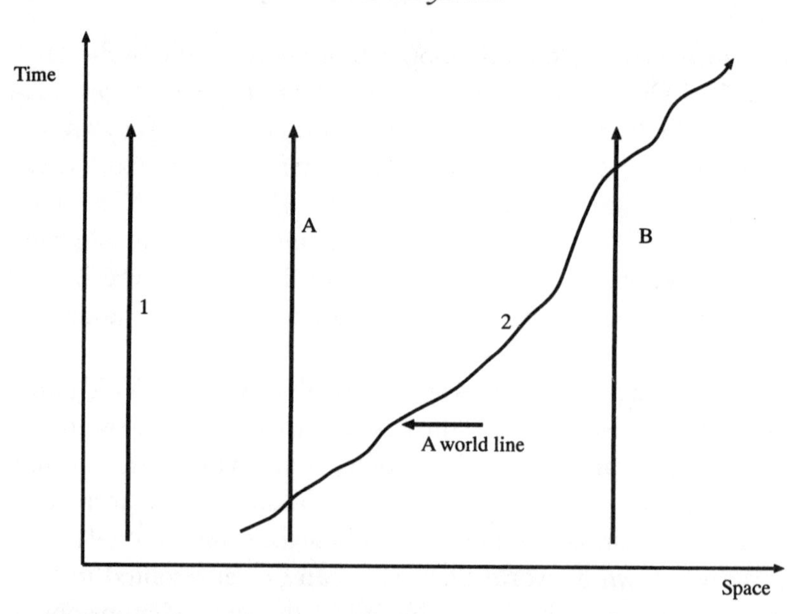

Figure 6. By representing time 'up the page' and space 'across the page' physicists can describe in simple geometric terms how particles move through spacetime. Particle 1 sits in one place, moving only in time (getting older). Particle 2 goes on a journey which takes it past the points A and B at different times. This is a spacetime diagram.

mathematicians, because (unlike dividing by zero) it is just the kind of thing that calculus is set up to handle, so that the equations give you a finite answer after you add up (integrate) the infinite number of infinitesimally small steps. 'At last', said Feynman, 'I had succeeded in representing quantum mechanics directly in terms of the action.' And although the representation had been set up using the idea of the wave equation as a guide, once the structure was in place that scaffolding could be removed without a trace, leaving a completely new description of quantum mechanics.

This new picture of the world was based on the idea of amplitudes. For each possible way that a particle can go from one point to another in spacetime there is a number, which Feynman called the amplitude. This number involves the action multiplied by a certain constant involving the mathematical i, the square root of -1, so it is called a complex number.

The important thing about a complex number is that it has two parts, which can be thought of in terms of little arrows. An arrow has a certain length, and it points in a certain direction. That's all there is to complex numbers, a length and a direction, which are kept apart by attaching this number *i* to one of them. Each little arrow represents a complex number. You add up the little arrows by placing them head to tail, but with each successive arrow turned through the appropriate angle (so it points in the right direction), and then drawing a new arrow from the tail of the first arrow to the head of the last arrow, which gives you both a length and a direction for the arrow that is the sum of all the little arrows. The direction of the arrow is also related to the phase of a wave. If you imagine the arrow attached to the spoke of a wheel, rotating around a fixed point, the direction in which the tip of the arrow points as the wheel turns goes up and down like a wave. When the arrow is pointing straight up, it corresponds to a peak in the wave. After the wheel has turned by 90 degrees, the arrow is horizontal. After another 90 degrees, it is pointing straight down, and that corresponds to a trough in the wave. After a further 90 degrees, it is horizontal again, and after the final 90 degrees needed to make up a circle it is back where it started. So the extent to which two waves are in phase with one another – how closely they match in step – can be described in terms of two little arrows, pointing in slightly (or considerably) different directions, or in terms of complex numbers.

The probability of a particle following a particular history is given by the square of the amplitude, and the probability of it getting from A to B at all is given by adding up all the amplitudes first, and then squaring the result.

In spite of Jehle's comments about the attitude of 'you Americans', Feynman did not immediately try to find a use for any of this. He did not apply his discoveries to practical problems in his thesis, but wrote down the general principles of his new approach to quantum mechanics, and developed the mathematical formalism. As we have seen, by this time he had plenty of other things on his mind, including both Arline's state of health and his war work, and at the time the thesis was written up, in the spring of 1942, his main concern was simply to get enough down on paper to satisfy the PhD examiners. Because of all these factors, it wouldn't be until 1948 that the path integral approach to quantum mechanics was published in the journal *Reviews of Modern Physics*, making it widely available to anyone who was interested. And it was only in the late 1940s, as well, that the new approach was triumphantly successful in solving the problems of quantum electrodynamics, as we discuss in Chapter 6. Perhaps for these reasons, the importance of Feynman's PhD thesis[11] itself is sometimes overlooked, and it's worth emphasizing its value here before we go on to look at all the events which helped to delay the completion of the theory of quantum electrodynamics.

One of the strange features of quantum mechanics is that right from the moment it was invented (or discovered) in the mid-1920s, there were two completely different descriptions of the quantum world. One was Schrödinger's approach, based on waves; the other was Heisenberg's approach, based on particles.[12] Both versions of quantum theory had been shown to be exactly equivalent to one another (by Dirac, among others), but most physicists worked with the wave equation (and still do), because it seemed comforting and familiar to people who had been brought up on wave equations. Now, Feynman had found a third approach to quantum mechanics, based on the action; arguably, that alone is enough to rank him with Schrödinger, Heisenberg and Dirac in the physicists' pantheon. It gave the same answers as the other two versions of the theory everywhere that they could be compared, and it could even handle problems that could not be solved using the wave function approach. It is also both relatively simple to use and clearly tied, through the Lagrangian, to the understanding of classical mechanics developed since the time of Newton. Wheeler has gone so far as to say that Feynman's PhD thesis marked the moment 'when quantum theory became simpler than classical theory'.[13]

This isn't just hindsight. In the same reminiscence, Wheeler tells how before Feynman had even completed his PhD Wheeler was visiting Einstein one day and couldn't resist telling him the news:

> Feynman has found a beautiful picture to understand the probability amplitude for a dynamical system to go from one specified configuration at one time to another specified configuration at a later time. He treats on a footing of absolute equality every conceivable history that leads from the initial state to the final one, no matter how crazy the motion in between. The contributions of these histories differ . . . in phase. And the phase is nothing but the classical action

integral, apart from the Dirac factor, $\hbar$. This prescription reproduces all of standard quantum theory. How could one ever want a simpler way to see what quantum theory is all about!

Indeed, as we shall see in Chapter 6, Feynman's path integral approach also works just as well in describing classical mechanics – so much so that Wheeler himself introduced Feynman's idea in the graduate course in classical mechanics that he was teaching that year. The point is not so much that quantum mechanics became simpler than classical mechanics, but that they became part of the same system – the same world view. Using Feynman's path integral approach, based on the Principle of Least Action, there is no longer any difference between classical mechanics and quantum mechanics, except for a trivial adjustment to the mathematics. Using the sum over histories approach, it is, in fact, possible to teach classical mechanics from the beginning (right back in school) in such a way that quantum mechanics follows on as a straightforward and logical development from familiar ideas.

But this approach never caught on. In universities around the world, even today, half a century after Feynman's insight, students are still taught classical mechanics the old-fashioned way, and then forced to train themselves into a new way of thinking in order to study quantum mechanics using the Hamiltonian approach and the Schrödinger equation. By the time most people learn about Feynman's approach (if they ever do), their brains have been battered by so much mechanics of one kind or another that it is hard to appreciate its simplicity, and galling to realize that they could have saved time and effort by learning quantum theory (and classical theory!) Feynman's way in the first place. Feynman's approach is not the standard way to teach physics for the same reason that the Betamax system is not the standard format for home video, and the Apple Macintosh is not the standard for personal computers, because an inferior system got established in the marketplace first, and continues to dominate as much through inertia and resistance to change as anything else. Such considerations, though, were hardly at the forefront of any physicist's mind in 1942, when Feynman finished his PhD, left Princeton and set off to work on the Manhattan Project in Los Alamos.

Notes

1. Told by Feynman to Mehra.
2. Richard Feynman, *Science*, volume 153, pp. 699–708, 1966. This is the published version of the Nobel lecture delivered in Stockholm on 11 December 1965. Hereafter referred to as 'Nobel lecture'.
3. Silvan Schweber mentions these earlier investigations of action at a distance in electrodynamics in *QED and the Men Who Made It* (see Bibliography).
4. Nobel lecture.
5. Nobel lecture.
6. *Surely You're Joking.*
7. Mehra.
8. Nobel lecture.
9. See Mehra. This conversation, the first between Dirac and Feynman, may seem a little terse. In fact, by Dirac's standards it was positively voluble. In 1929, just 27 years old but already an acclaimed genius who had made a major contribution to the development of quantum theory, he visited the University of Wisconsin. The *Wisconsin State Journal* published an interview with the young genius in which his part of the conversation consisted almost entirely of monosyllables. In a typical exchange, the reporter asks if Dirac ever goes to the movies. 'Yes', he replies. 'When?', the interviewer asks. 'In 1920.'
10. Nobel lecture.
11. Richard Feynman, *The Principle of Least Action in Quantum Mechanics*, PhD thesis, Princeton University, May 1942.
12. See, for example, John Gribbin, *In Search of Schrödinger's Cat.*
13. *Most of the Good Stuff.*

— 5 —

From Los Alamos to Cornell

Once a decision had been made to build an atomic bomb, the first problem that had to be tackled was getting enough radioactive material of the right – potentially explosive – kind to do the job. The runaway process of nuclear fission that powers such a bomb happens when nuclei of a certain kind of heavy element split into two or more lighter nuclei, with energy being released. The key to an explosive chain reaction is that when one nucleus splits it also releases two or more neutrons, which collide with other nuclei and make them split in turn, in a growing cascade. Calculations showed that an explosive chain reaction ought to occur if a sufficiently large quantity of either of two radioactive elements, uranium-235 or plutonium-239, could be brought together in the right way in a bomb.

Plutonium could only be manufactured artificially, by the bombardment of another, more stable form of uranium, uranium-238, with subatomic particles. The difference between uranium-235 and uranium-238 is simply that each nucleus of uranium-238 contains three more neutrons than each nucleus of uranium-235, but that is enough to make the nucleus relatively stable – it is radioactive, but has a very long lifetime. Uranium-235 is much more radioactive and potentially explosive, but occurs naturally mixed with uranium-238 in trace quantities, with only 7 atoms of uranium-235 for every 1000 atoms of uranium-238. Both routes to a nuclear bomb were taken by the Manhattan Project: the manufacture of plutonium-239, and the separation of uranium-235 from naturally occurring uranium.

The Princeton project that Feynman joined under Robert Wilson, before completing his thesis, was one of several attempts to find a way of separating out uranium-235 in the quantities required.

Progress was slow, which was partly why Feynman was, after all, able to take time off in the spring and early summer of 1942 to complete his PhD and get married. Late in 1942, the Princeton approach to the uranium separation problem was abandoned, in favour of a technique being developed at Berkeley, in California, which was making more rapid progress. But the entire Princeton team, along with other researchers involved in the atomic bomb project, was invited to move to Los Alamos, where a new, secret research centre was being built to solve the problem of actually building the bomb and making it work. They all signed up, but then had to sit around, twiddling their thumbs for several months while the Los Alamos lab was actually being built.

In order to make good use of the time, Wilson found several minor problems for the team to tackle. He also sent Feynman off to visit the Metallurgical Laboratory in Chicago, at that time the heart of the Manhattan Project, where Enrico Fermi's team was building the world's first nuclear reactor, then called an atomic pile. Wilson wanted as much information as Feynman could get about the whole Top Secret project. Feynman's reminiscence about his war work can be found in an article called 'Los Alamos from Below';[1] in that article, he recounts how Wilson instructed Feynman to go to each group of researchers in Chicago, say that he was going to be working with them, and ask for enough information about the project to enable him to start work. His conscience troubled him, because he expected to go away from Chicago without having given anything back in return. But, as Wilson may well have expected when choosing Feynman for the job, it didn't work out like that. He not only obtained all the information that Wilson wanted, but everywhere he went he made valuable suggestions, helping the work in Chicago along. In an obituary of Feynman published in 1988,[2] Philip Morrison, a member of Fermi's Chicago team, recalled how 'we all came to meet this brash champion', who 'did not disappoint us; he explained on the spot how to gain a quick result that had evaded one of our clever calculators for a month'. The way Feynman told the story, it was just luck that he happened to know a mathematical trick that would work on that problem; to everybody else, it was an example not only of his mathematical genius, but his ability to see to the heart of a problem as soon as it was suggested to him. Both abilities would shortly be exercised to the fullest.

The scientific head of the Manhattan Project was Robert Oppenheimer, a renowned physicist whose work before the war included investigations involving the Dirac equation and theoretical studies of what are now known as neutron stars and black holes, three decades before such objects were discovered. Like Feynman, Oppenheimer would eventually (in 1967) die of cancer. Suggestions that this was possibly related to his wartime work with radioactive materials should be taken with a pinch of salt; Oppenheimer's cancer of the throat was more probably a result of his chain-smoking of cigarettes.

Oppenheimer was the ideal man to head the scientific side of the Manhattan Project, providing an interface between the scientists and the military, having a thorough understanding of the science involved in the project and, not least, taking a close personal interest in the wellbeing of everyone at Los Alamos. In Feynman's case, this extended to finding a sanatorium for Arline to stay at in Albuquerque, as close as possible to Los Alamos itself. Feynman was deeply touched by this personal attention, and like the rest of the team would do anything for 'Oppie'.

The nearest railway station to Los Alamos was at Lamy, New Mexico, and when the Princeton group were finally given the OK to move out to New Mexico, that was where they were headed. But Princeton was a small town, and the authorities were worried that curiosity might be aroused if all the physicists suddenly upped and left for Lamy. So they were told to buy their train tickets elsewhere, in order not to cause gossip locally. Feynman, as ever, had his own way of looking at things. If everybody else was buying their tickets in other towns, it would do no harm if one person – Feynman – bought a ticket to Lamy at Princeton station. So he did. 'Oh,' said the man at the ticket office, 'so all this stuff is for you!' The team had been shipping out crates of material for weeks, all going from Princeton to Lamy by rail. At least, by buying his ticket in Princeton, Feynman had explained who all the stuff was for.

Feynman was one of the first scientists to set out for Los Alamos, leaving Princeton, with Arline, on 28 March 1943. They paid extra for a private suite on the train, treating the long cross-country ride as a holiday. Arline had hoped that it might also be something of a honeymoon – as yet, the marriage had not been consummated, partly through lack of opportunity, partly through fear of the effect on Arline's health, and the possibility of Richard's becoming infected with TB. But it seems that nothing came of these hopes.[3]

Feynman's own accounts of his time at Los Alamos focus on the fun and games – his safecracking exploits, and the battles with the censor, being classic examples. The censorship problems arose because both Arline and Melville used to write to Richard in code, using systems they had invented, and which he didn't have a key to. The game – light relief in the midst of his serious work on the project – was for him to crack the code in order to read the letters. But the censors wouldn't allow coded messages to go in or out of Los Alamos! The situation was eventually resolved when the censor agreed to allow this, provided a key was included in each letter so that the censor could read it before passing it on, without the key, to Feynman. Feynman milked the deliciously bureaucratic problem for all it was worth. The very fact that the mail was being censored was officially a secret, so when Feynman was told to tell Arline not to mention censorship in her letters, he promptly wrote her a letter beginning, 'I have been instructed to inform you not to mention censorship in your letters.' The censor, of course, censored the letter; Richard had to go and tell Arline what was going on in person.

The fun and games may seem childish, but they were an important safety valve for Feynman and his colleagues – more so for him than for most. He was not quite 25 when he arrived in Los Alamos, where he had another piece of what he called luck. It happened that most of the big shots were away at the time, and Hans Bethe, who was the head of the Theory Division on the project, needed someone to bounce some ideas off. He came into Feynman's office, and started talking physics. As always, when talking physics Feynman forgot who he was talking with, and was unimpressed by their status. He told Bethe his ideas were crazy, Bethe argued his case, Feynman responded by pointing out the flaws in it, and the arguing went on until the problem was solved, just as in the debates with Wheeler back in Princeton. Bethe, who already knew of Feynman's reputation, was impressed, and made him a group leader in the Division, in charge of a team of four other researchers. He was the youngest group leader by about 10 years (Bethe himself was 37 in 1943), and showed a real flair for getting the best out of his team.

Feynman became something of a troubleshooter, a Mr Fixit, for the whole Theory Division. Always fascinated by mechanical things, he spent a lot of time repairing mechanical calculators (glorified adding machines) and typewriters, until Bethe decided that this was a waste of his talents and ordered him to stop.[4] But then the project took delivery of a new kind of calculating machine, delivered from IBM in many boxes. Feynman and a colleague took the parts out of the boxes and put them together to assemble the machines. A week later, the official IBM engineer, drafted for the duration, arrived to assemble and look after the machines; he told Bethe that he had never seen such machines put together by non-experts before, but that they were all working properly.

When the machines were first put to use, however, there were problems. The leaders of the group using the calculators were fascinated by their abilities, and loved to play with them, but the real work, calculating important numbers needed for the construction of the bomb, simply wasn't getting done. Bethe's answer was to put Feynman in charge of the IBM machines, as head of the Theoretical Computations Group – by then, the most important group in the Theory Division. 'Soon the group was working efficiently, and we got our answers promptly and steadily.'[5]

Feynman's ability was also noticed by outsiders. When Niels Bohr, one of the greatest physicists and a founding father of quantum theory, visited the project, he noticed the way Feynman spoke up in meetings, cutting to the heart of the problem in question. The next time Bohr came to Los Alamos, Feynman received a call from Bohr's son, also a physicist, asking him to meet the great man early, at 8 a.m., before the big meeting. For a couple of hours Bohr went over his ideas, with Feynman, as usual, interrupting to point out what was wrong with them, shouting, 'you're crazy', treating Bohr the way he would treat any physicist. At last everything was sorted out. 'Well,' said Bohr, 'I guess we can call in the big shots *now*.' The younger Bohr explained – after the previous visit, Bohr senior had commented on Feynman's contributions to the discussion. 'He's the only guy who's not afraid of me, and will say when I've got a crazy idea. So *next* time when we want to discuss ideas, we're not going to be able to do it with these guys who say everything is yes, yes, Dr Bohr. Get that guy and we'll talk to him first.'

With all this going on, Richard still had Arline to consider. Every week, he would hitchhike (or if he was lucky, borrow a car) to travel almost 100 miles to see her in Albuquerque on the Saturday afternoon. He would stay in a cheap hotel overnight, visit her again on Sunday morning, and make his way back to Los Alamos in the afternoon. He used the long trips to think about quantum mechanics, developing the ideas in his thesis further. Given the pressures he was under, it's hardly surprising that when the opportunity presented itself he couldn't resist winding up the censors, or having fun in whatever other way he could.

Much of that fun was provided by Arline, sometimes to Richard's embarrassment. In *What Do You Care*, he tells how she reminded him of his own dictum, to take no notice of what other people thought. The sanatorium was right on Route 66, the main road across the United States, with trucks rolling by what passed for the lawn in front of the building. As part of her attempt to lead a normal life, Arline bought, by mail order, a little barbecue grill, and made Dick go out on the lawn and cook her steaks, most weekends, dressed in a chef's hat and apron. At first he protested. But – 'What do *you* care what other people think?', she responded. The first Christmas in New Mexico, Arline ordered a batch of printed cards for the couple to send out, nice cards with the message inside 'Merry Christmas, from Rich & Putsy'. Feynman protested that they were too informal to send to important people like Oppenheimer and Bethe, but out they went anyway. The next year, by which time Feynman was on familiar terms even with the senior scientists like Bethe, she showed him another batch of cards, with the message 'Merry Christmas and a Happy New Year from Richard and Arline Feynman'. As soon as he expressed his relief at their suitability, she produced another box of cards, especially for the important people, signed 'From Dr & Mrs R. P. Feynman'. Out they went, resulting in Feynman being ribbed by his colleagues about the stuffy formality of his Christmas greetings.[6]

Of course, his friends knew it was a joke. Many of them visited Arline in the sanatorium – even Wheeler, on a visit to Los Alamos, found time to call in on her – and they were pleased with anything that made her happy. Always busy, she taught herself Chinese calligraphy, and made many plans for a future in which Dick would be a real professor and they would raise a son called Donald. In *What Do You Care*, Feynman explained that neither of them felt overwhelmed by her condition, that they had 'a hell of a good time together'. After all, he points out, everybody knows that they will die eventually. The only difference for them was that they had five years together instead of 50 years:

> Why make yourself miserable saying things like, 'Why do we have such bad luck? What has God done to us? What have we done to deserve this?' – all of which, if you understand reality and take it completely into your heart, are irrelevant and unsolvable. They are just things that nobody can know. Your situation is just an accident of life.[7]

Back at Los Alamos, Feynman was making another reputation for himself, as a teacher. Part of the reason for his success in getting the Theoretical Computations Group working so well was because he explained to them what was going on. The people operating the machines were essentially kids fresh out of high school, called up and dumped in barracks, told to operate these machines using punch cards, without any clue as to what the work was all about. It needed Oppenheimer to arrange a special security clearance, but then Feynman told them all about the project and how important their work was, firing them with enthusiasm. In the nine months before he took over the group, it had solved three problems. In the three months after he took over, it solved nine problems – the same people, using the same machines, but under new leadership.

The uranium for the bomb was actually being separated at Oak Ridge, Tennessee. Again, the industry workers in the plant where the work was being done didn't know what the work was for. And, again, progress was slow and difficult. Emilio Segre, one of the Los Alamos team, was eventually sent to Oak Ridge to identify some of the problems, and while carrying out a preliminary inspection of the plant he was horrified to find large amounts of unpurified uranium nitrate being kept in solution in large tanks. If the pure uranium-235 were to be stored in the same way, it would explode. The military people in charge at Oak Ridge knew that a certain amount of pure uranium-235 (the so-called critical mass) was needed to cause an explosion, but they hadn't realized that when neutrons are slowed down by passing through water they are much more effective at causing fission. Considerably less uranium-235 in a solution of this kind would still be a hazard.

Segre brought back all the information he could glean about how uranium was being refined and stored at Oak Ridge. The Los Alamos scientists studied the information and worked out appropriate safety procedures. Then, somebody had to go and explain it all to the workers at Oak Ridge. Who else but Dick Feynman? Before he left, Oppenheimer told him how to make sure he was given a hearing. If there were any objections on grounds of security, he had to say, 'Los Alamos cannot accept the responsibility for the safety of the Oak Ridge plant unless . . .' The mantra worked like a charm. He explained everything, about fission, the role of neutrons, how they behave when they pass through different substances and so on. The plant was redesigned to avoid the possibility of too much purified uranium-235 piling up in one place, and as a side effect the workforce became much more enthusiastic about the project and worked much more efficiently. Many of the people involved felt that Feynman, on behalf of the Los Alamos team, had prevented a disastrous accident and saved their lives.

Not that the Los Alamos team themselves weren't, by modern standards, horrifyingly careless of their own safety when handling radioactive material. To be sure, they took care to avoid the build-up of a critical mass of either uranium-235 or plutonium as the material became available. But as the first atomic bombs were being assembled, these highly radioactive materials were handled with few of the safety precautions that would now be *de rigueur*. Of course, it was wartime, and at that time the risks of radiation poisoning were poorly understood. But apart from the, perhaps necessary, risks involved in handling the radioactive material, the team kept, in one of the rooms at Los Alamos, a small silver-plated ball mounted on a pedestal, something to impress visitors with. The ball was made of plutonium, today regarded as one of the most toxic substances on Earth. It was warm to the touch, because of its radioactivity – warmer, at the high altitude of Los Alamos, than it would have been at sea level, because cosmic rays striking the nuclei inside the ball triggered additional fission reactions. With hindsight, it would have been surprising if some members of the team had *not* eventually died of cancer.

With the benefit of modern knowledge, perhaps Feynman and his colleagues could have taken precautions that would have extended their own lives. But they didn't have that knowledge, and their situation was, in his own words, 'just an accident of life'. Equally, there was nothing he could do to save Arline's life. By the beginning of 1945, many factors in Feynman's life were coming to a head. The Manhattan Project itself was nearing completion. Meanwhile, the cost of keeping Arline in hospital was beginning to be a problem. In a letter to 'Dearest Putzie' dated 24 April 1945, Richard spelled out their financial situation. His income was $300 a month; after meeting his own modest expenses and her hospital bills, they needed another $300 a month, which was coming from Arline's dwindling savings, now down to $3300. They could keep going at this rate for another 10 months, but Richard asked whether it might be 'necessary to sell the ring and piano now'.[8] He also offered to go back to eating in the mess hall, saving $15 a month. But Arline was wasting away, and realistically Feynman cannot have expected her to last long enough for them to run out of money.

Against this background, they at last, at Arline's instigation, made love. It was a last stand against the inevitable, perhaps Arline's desperate attempt to leave Richard with the child that they both yearned for, even if she could not stay with him herself. She missed her next period, and was overjoyed at the prospect of being pregnant. But she was not; it was just another symptom of her illness.

Her condition continued to deteriorate, so much so that she asked Richard not to visit her on some weekends. In May, her father made the long and difficult wartime journey from New York to see her for the last time. One day in June he called Feynman at Los Alamos, and told him the end was near. Borrowing a car from Klaus Fuchs, Richard made it to Albuquerque in time to be with her when she died. The next day, he returned to Los Alamos and buried himself in his work, not grieving properly until, months later, he was in Oak Ridge and noticed a pretty dress in a shop window: 'I thought, "Arline would like that," and then it hit me.'[9]

Soon after Arline's death, Feynman was able to take a short break in Far Rockaway as the project neared completion. He was there when he received a message from Bethe, saying, 'The baby is expected.' He flew back to New Mexico just in time to be present at the Trinity test, where he was part of the group of observers 20 miles from ground zero. Everyone had been issued with dark glasses to protect their eyes from the ultraviolet radiation produced in the explosion; Feynman, still his own man, knew that even ordinary glass stops ultraviolet light, and calculated that the ordinary light from the explosion wouldn't be bright enough to damage his eyes. So he watched the explosion through the windshield of a truck – the only person to watch the first nuclear explosion on Earth with his naked eyes.

The immediate reaction among the team was euphoria that their work had been successful. Only much later did people start to ask the questions which now concern historians, about the morality of proceeding with the Manhattan Project once it was clear that Germany was being defeated and it was known that there was no nuclear threat from Japan, and whether the Hiroshima and Nagasaki bombs should ever have been dropped. Feynman's own counter-reaction was both more immediate, and more personal. By the end of the year, when he was still just 27 years old, he was teaching at Cornell University, in Ithaca, New York; he recalls sitting in a restaurant in New York City around that time, working out how much of the city would be destroyed by a bomb the size of the one dropped on Hiroshima:

> I would go along and I would see people building a bridge, or they'd be making a new road, and I thought, they're *crazy*, they just don't understand, they don't *understand*. Why are they making new things? It's so useless.[10]

Happily, Feynman's assumption of the inevitability of nuclear war has, so far, proved wrong. But the story gives a good idea of his state of mind in his early years as a 'real professor' at Cornell.

He had chosen Cornell because that was where Bethe worked. By 1945, Feynman's reputation was such that he could have had his pick of several posts (although he didn't seem to be fully aware of his 'market value'); but he had got on well with Bethe at Los Alamos, and had been impressed by Bethe's skills both as a physicist and as a mathematician who knew even more mathematical tricks and short cuts than Feynman did. Technically, Feynman's first academic position was with the University of Wisconsin – after completing his PhD, while still working in Princeton on the bomb project, he had accepted the offer of a job there, without pay, suspended until the war work was finished. But he never took up the post. By the end of October 1943, Bethe was already urging Cornell to sign Feynman up, and this led to the offer of an appointment from the autumn of 1944, with a leave of absence (again, unpaid) granted for the duration of the war. Feynman was happy to accept, and later said that he didn't consider the other offers he received, because he wanted to be with Bethe.

But that didn't stop the other offers coming in. One of the most frustrated players in the story was Oppenheimer, who wanted to lure Feynman to his own home base, at the University of California, Berkeley. His correspondence[11] shows how strongly he felt about this. In a letter to Raymond Birge, the head of Berkeley's physics department, dated 4 November 1943, Oppenheimer describes Feynman as:

> The most brilliant young physicist here, and everyone knows this. He is a man of thoroughly engaging character and personality, extremely clear, extremely normal in all respects, and an excellent teacher with a warm feeling for physics in all its aspects . . . Bethe has said that he would rather lose any two other men than Feynman from this present job, and Wigner said, 'He is a second Dirac, only this time human.'

It wasn't enough to persuade Berkeley to make an immediate offer to Feynman, and six months later, on 26 May 1944, Oppenheimer was still banging his head against their bureaucratic brick wall:

> It is not an unusual thing for Universities to make commitments to young men whom they wish to have after the war . . . [Feynman] is not only an extremely brilliant theorist, but a man of the greatest robustness, responsibility and warmth, a brilliant and lucid teacher, and an untiring worker. He would come to the teaching of physics with both a rare talent and a rare enthusiasm . . . he is just such a man as we have long needed in Berkeley to contribute to the unity of the department and to give it technical strength where it has been lacking in the past.

Eventually, Berkeley did make Feynman an offer, but he turned it down, happy to be going to Cornell. The authorities at Cornell, though, didn't know that Feynman had no intention of going anywhere else, and kept hearing, via Bethe, rumours about the offers he was getting from other universities, including Berkeley. As a result, from time to time, while he was still at Los Alamos, Feynman received notification that he had been awarded a raise in his notional salary. By the time he actually arrived there, and started drawing the salary, it had been increased to $3900 a year, a very healthy rate for the job in 1945, with every prospect of reaching the $5000 per annum that the young Feynman had always hoped for.[12]

Feynman's efforts to settle down as a 'dignified professor' were doomed to failure, as he recounts in *Surely You're Joking*. He left Los Alamos earlier than most of the physicists, and arrived in Cornell to take up his appointment at the beginning of November 1945, having worked out on the train an outline of the course he was to teach. He had no trouble with the teaching side of being a professor; it was the 'dignified' bit that somehow never seemed to work. For a start, without Arline and the home they had both longed for, he preferred the community life on campus to a solitary apartment. He lived much the same kind of life as he had as a graduate student at Princeton and undergraduate at MIT, only now he had a fund of anecdotes about his wartime experiences with which to regale people at dinner, beginning the development of the 'colourful character' persona. Still young, and looking younger, he tried to make a new social life without Arline by going to student dances, where he was puzzled by his initial lack of success with the ladies. It turned out that they regarded his matter of fact claims to be a professor of physics who had worked on the atomic bomb project as an outrageous line-shooting exercise, and he got on much better when he said nothing about his war work and allowed them to think he was a freshman on the GI Bill, for soldiers returning from the war.

Underneath it all though, Feynman was, by his own standards, lonely and depressed. Nobody noticed. Many years later, Bethe explained – 'Feynman depressed is just a little more cheerful than any other person when he is exuberant.'[13] His depression was understandable – the death of Arline, and the end of the stressful years of war work were beginning to catch up with him. On top of that, after a few months at Cornell Feynman began to worry that he was burned out. He thought he wouldn't be able to think about fundamental physics any more. Then, on 7 October 1946, Melville (who had long suffered from high blood pressure) had a stroke; he died the next day. Shortly after the funeral, Feynman wrote a last letter to Arline, which he never showed to anyone and which was found among his papers after his own death. It told her how much he loved her, and how empty his life still seemed without her. He ended with a poignant P.S.: 'Please excuse my not mailing this – but I don't know your new address.'[14]

Against this background of inner turmoil and the conviction that he was burned out, Feynman continued to receive offers of posts at other universities and to get raises in salary as a result. A few months after writing his last letter to Arline, early in 1947, he received the offer to end all offers. It was from the Institute for Advanced Study in Princeton. Knowing that Feynman felt that the Institute was too 'theoretical', an ivory tower cut off from the hurly-burly of a normal university, they offered to create a special post just for him, so that he could spend half his time at the Institute and half at Princeton University. It was a dream position, a position, in Feynman's words, 'better than Einstein's'. And the salary was impressive, too. All in all, he wrote in *Surely You're Joking*, 'it was ideal; it was perfect; it was absurd!'

On the spot, he decided that the whole business was ridiculous. Nobody could live up to the expectations others had of him. He certainly couldn't, so he wasn't going to try any more.

The very same day, perhaps because he had overheard Feynman talking along these lines to his colleagues, Robert Wilson, who was by now head of the Nuclear Research Laboratory at Cornell, called Feynman into his office, and told him not to worry about research. As Feynman later paraphrased it, he said, 'You're teaching your classes well; you're doing a good job, and we're satisfied. Any other expectations we might have are a matter of luck. When we hire a professor, we're taking all the risks. If it comes out good, all right. If it doesn't, too bad. But you shouldn't worry about what you're doing or not doing.'[15]

So Feynman was officially freed from the responsibility of coming up with any brilliant new ideas. He had said farewell to his father, and written his last letter to Arline. In the spring of 1947, he remembered how he used to enjoy doing physics – how it used to be fun, not a chore. He decided that he had a cushy job, secure for life, teaching classes that he rather enjoyed. He wouldn't look for any more big problems to solve; instead he would play with physics, for fun, the way he used to.

A few days later he was in the cafeteria when one of the students, fooling around, threw a plate into the air, spinning it like a modern frisbee. Like all the plates, it had the red medallion of Cornell on it, and Feynman noticed that as the plate wobbled and spun the medallion went round at a different rate from the wobble. Intrigued, and just for fun, he set out to calculate the relationship between the wobble and the spin, and found that there is a precise ratio, 2:1, which comes out of a complicated equation.* He told Bethe the news. Bethe asked why he had done the work. For fun, Feynman replied; there's no importance in it at all.

But he was wrong. As his subconscious may have been well aware all along, the big problem that he was stuck with in developing his thesis work was how to include the effects of the spin of the electron in these calculations. The equations that Feynman had played with in calculating the wobble of a spinning plate were directly relevant to that problem. As he realized this, he slipped easily into fresh work on the old problem. 'It was', he said in *Surely You're Joking*, 'like uncorking a bottle. Everything flowed out effortlessly.' Physics was fun again, and 'The whole business I got the Nobel Prize for came from that piddling around with the wobbling plate.'

Well, it wasn't really quite that easy, or quite that quick. Feynman's route to the work that won him the Nobel Prize, from the spring of 1947 onwards, can actually be marked out by the events associated with three select gatherings of scientists, each organized by Oppenheimer on behalf of the National Academy of Sciences, in 1947 and the next two years.

The first of these meetings took place from 2 to 4 June 1947, at

*In *Surely You're Joking* Feynman gave the ratio in the wrong way round, 2:1 for spin:wobble instead for wobble:spin, and the Cornell medallion on the plates may really have been blue, not red. Like all Feynman anecdotes, the precise details don't matter, but the moral is clear.

the Ram's Head Inn on Shelter Island, right at the tip of Long Island. The official theme of the meeting was 'Problems of Quantum Mechanics and the Electron', but it has gone down in scientific history simply as 'the Shelter Island Conference'. It was Feynman's first opportunity to participate in a scientific gathering with some of the top physicists in peacetime, and, with just 24 participants in all, the gathering was small enough to get real work done, in a manner reminiscent of some of the brainstorming sessions of the Manhattan Project. Apart from Feynman, the other bright young man at this gathering of the great was Julian Schwinger, a professor at Harvard University. Schwinger was an almost exact contemporary of Feynman (he had been born three months earlier, on 12 February 1918, also in New York City), and a renowned prodigy, already with a string of papers to his name. He had actually completed the work that became his PhD thesis *before* he graduated (from Columbia University) in 1936, at the age of 18.

The big talking point at the Shelter Island Conference was an experimental discovery that had been made a few weeks before, at the end of April, by Willis Lamb and his colleague Robert Retherford at Columbia University. They had been probing hydrogen atoms using beams of microwaves – a technique developed directly from Lamb's own war work, on radar – to measure the energy levels of the electrons in those atoms. In effect, they were measuring the spacing between the rungs on the energy level ladder. According to the Dirac theory, the electron in a hydrogen atom could exist in either of two quantum states which had precisely the same energy, as if there were a double rung on the ladder. But Lamb found that one of these states had slightly more energy than the Dirac theory predicted, so that there was a tiny separation between the two energy levels. One of the energy levels was shifted slightly – one rung of the pair on the ladder was slightly higher than its companion. This became known as the Lamb shift. The Shelter Island Conference got all this from the horse's mouth, because Lamb was one of the participants. An almost equally dramatic discovery, a precise measurement of the magnetic moment of the electron, was reported to the meeting by Isidor Rabi, but was overshadowed by Lamb's work; soon, though, (as we shall see in Chapter 6) it also played a major part in the development of quantum electrodynamics.

In one sense, the discovery of the Lamb shift was an indication that the Dirac theory was incomplete. But physicists already knew that, because of the way infinities came into the theory of quantum electrodynamics (QED) when they tried to calculate the self-interaction of an electron in an electromagnetic field. Indeed, the infinite term resulting from the self-interaction would, if it were real, have corresponded to an infinite 'Lamb shift', whatever that might mean. So in another sense, Lamb's work showed that the Dirac theory might not be so bad after all, because the disagreement with the experiments, far from being infinite, was a tiny number corresponding to a very small shift in the energy levels. If Lamb had found zero shift, that would have meant Dirac was right, which would have flown in the face of what was already known, and would in that sense have been bad news. But the Lamb shift told the physicists at Shelter Island that what they had to try to find was not zero or infinity but a finite, very small, and now precisely known, quantity. That, they thought, they ought to be able to handle; with real numbers on the table in front of them, perhaps there was, at last, a chance to make sense out of QED.

Like other participants, Feynman also contributed to the conference, a talk about his spacetime approach to quantum mechanics, and path integrals; but, like most of the other presentations, this contribution (essentially a summary of his thesis work) made very little impact alongside the sensational news about the Lamb shift. The big question was, could quantum theory be tweaked up to predict the right amount of change in the energy levels?

At that time, Hans Bethe had a summer job as a consultant for the General Electric Company's research laboratory in Schenectady, New York, and it was on the train from New York to Schenectady, immediately after the Shelter Island Conference, that he made the first, imperfect but suggestive, calculation of the Lamb shift. Bethe seemed to like working on trains – in similar circumstances, back in 1938, he had solved the puzzle of how nuclear fusion reactions keep the Sun hot (the work for which he won his own Nobel Prize) on a train ride back to Cornell after a conference in Washington D.C. Now, he had worked out a trick to get rid of the infinities in QED, and leave behind a small, finite amount of interaction, corresponding to the Lamb shift. There was one snag; in this first stab at the problem, he had not taken account of the effects required by relativity theory, and only had a non-relativistic calculation of the shift. But still, it was a big step in the right direction.

What Bethe did, in effect, was to calculate the energy for an electron in a hydrogen atom, which came out as the usual infinity plus a correction caused by the presence of the nearby atomic nucleus (in this case, a single proton). Then he subtracted from this the energy of a free electron, which is infinity, leaving behind the correction – the energy shift required. This approach, called 'renormalization', came originally from work by the Dutch physicist Hendrik Kramers (another of the Shelter Island participants) on another puzzling infinity that arises in quantum theory, and it has no right to work. Infinity is a funny thing. Infinity plus a little bit is also infinity, and at one level you might think that subtracting the two quantities Bethe was playing with (infinity plus a little bit, minus infinity) ought to leave zero. On the other hand, you could imagine 'making' infinity by adding up all of the integer numbers there are, and making another infinity by doubling each integer and adding up the doubled numbers. Bizarrely, the second infinity is *smaller* than the first one, because it contains only even numbers, whereas the first infinity contains all the even numbers *and* all the odd ones. If you subtract the second infinity from the first one, now you will be left with infinity again, the sum of all the odd numbers alone! In fact, a mathematician can arrange to get almost any answer you want by subtracting infinity from infinity; the fact that, as Bethe found, the infinities really could be cancelled out of the quantum equations in this way to give the right answer for the Lamb shift seemed to some people a miracle, to others a fraud, while to most physicists it meant that he had made a fundamental discovery about the way the world works, although they weren't quite sure what that discovery was (this final position is still roughly where physics stands today).

The discovery highlights one of the great features of Bethe's work. Given a number, a link with experiment, he would take the appropriate theory and shake it by the scruff of its neck until either it fell apart or it was forced to agree with the experiment. Feynman's great weakness, up to this point, was that he had developed a whole new way of looking at quantum theory, but had never tried to use it to calculate numbers that could be compared with experiment in this way. He still had not learned the lesson of his encounter with Jehle. And yet, one of the great features of Feynman's version of quantum theory was that it had relativity built into it – it was, in the jargon, relativistically invariant. As news of Bethe's work spread, many physicists tried to find a way to develop a relativistic version of the appropriate equations. Feynman first heard the news in an excited phone call from Bethe in Schenectady, but didn't immediately take in its importance.[16] It was only when Bethe returned to Cornell and gave a formal lecture on his discovery, ending by pointing out the need for a relativistically invariant version of the calculation, that the penny dropped. Feynman went up to Bethe after the lecture and said, 'I can do that for you. I'll bring it in for you tomorrow.'[17]

Up to that point, though, Feynman hadn't even used his beautiful new machinery to calculate the self-energy of the electron. For the first time,[18] he applied the path integral approach to ordinary electrodynamics, instead of using the half advanced and half retarded formulation. The theory was clear enough, but Feynman had never tried to do anything like this with it before. Probably as a result, when he did try to work through the Lamb shift problem with Bethe the next day, somehow he made an error, and when they tried to apply renormalization the infinities refused to disappear (in other words, the equations diverged). He had to go back to his room and worry away at the problem, learning how to calculate the self-energy and all the other things he had ignored, over the next couple of months. Then he tackled the problem again. The calculation worked out right, and the infinities disappeared – in the jargon, the equations converged – in just the right way, using the renormalization trick. It was now early in the autumn of 1947. Having realized, at last, the power of his new tool, Feynman set out to calculate everything in sight. By the time of the next of the three big meetings, the Pocono Conference held in April 1948, he had done just about all of the work for which he would win the Nobel Prize, including an updated discussion of positrons as electrons going backwards in time; but the material was not yet in a form that could be immediately understood by other physicists, brought up on the old ways of the Hamiltonian approach and the Schrödinger equation.

Some of Feynman's new work was presented in a talk he gave at the Institute for Advanced Study, in Princeton, on 12 November 1947. Dirac was in the audience, and one of the other participants wrote to a colleague that 'Dirac is very impressed by Feynman, and thinks he does some interesting things.'[19] But Dirac was then in a minority, as far as appreciating Feynman's new work was concerned.

For most physicists, the next really exciting development in QED came from Julian Schwinger, who presented his version of the Lamb shift calculation, in relativistically invariant form, to the annual meeting of the American Physical Society which took place in New York in January 1948. He had also calculated the important property called the magnetic moment of the electron, and the extent of its departure from the value predicted by the Dirac equation. So many people wanted to hear the talk that it had to be repeated in the afternoon. After this talk, Feynman, who was in the audience, stood up and mentioned that he, too, had got the same results (in one case, going a step further than Schwinger) by a different method. He later regretted this. Schwinger was, at the time, more well known than Feynman (not least because Feynman had hardly published anything since his undergraduate senior thesis; even the work in his PhD thesis would only be published in a journal, *Reviews of Modern Physics*, in 1948), and Feynman felt that his comments came across with the air of a small boy saying 'me too', when he had really just been trying to say that the results must be right if two separate calculations gave the same answers.[20] In Feynman's own mind, though, it was an important moment, because it meant that he really was on the right track, if he was getting the same results as Schwinger. Of course, there was an element of rivalry, felt especially keenly by Feynman because he was the lesser known of the two. He wanted to catch up with Schwinger, and overtake him; but most of all, he wanted to solve the problems of QED, whether Schwinger solved them first or not, just as, long ago, he had solved mathematical problems for his own satisfaction, without worrying whether some Greek mathematician in ancient times had solved them first.

The trouble with Schwinger, though, was that his work was difficult to follow because it was so complicated. This was partly in the nature of the Hamiltonian approach, and partly, many physicists suspected, through Schwinger's own love of mathematics. If there were two ways to prove a mathematical point, it seemed, Schwinger would always choose the more elegant but also more complicated way, showing off his erudition in the process. It meant that his version of quantum electrodynamics involved hundreds of equations, developed with great mathematical skill and precision, but with few signposts, in the form of links with physics of the kind Bethe so revelled in, to point the way. Schwinger was a virtuoso with equations, but to anyone lacking his virtuosity it was often hard to fathom where he got his answers from. Nevertheless, his great triumph – the last great fling of the old way of doing quantum mechanics – came at the conference held at the Pocono Manor Inn, in the Pocono Mountains of Pennsylvania, between 30 March and 2 April 1948.

This time, there were 28 physicists at the meeting. Schwinger offered them their first glimpse of a complete relativistically invariant theory of quantum electrodynamics, taking up almost a full day. There were few questions, because nobody there had enough mathematical skill to find any flaws in the argument, even if there had been any. But everyone agreed that it was a triumph. Then Feynman, seven weeks short of his thirtieth birthday, gave his talk, under the title 'Alternative Formulation of Quantum Electrodynamics'. Partly at the suggestion of Bethe, who had noticed how Schwinger's equations stunned the audience into silence, he made the mistake of offering his version of QED from a mathematical perspective, instead of kicking off from the physics he knew and loved so well. Feynman's approach was new and unfamiliar, and nobody understood it. When he talked of electrons going forwards and backwards in time, they were baffled. There was no communication. In the end, he gave up. He knew he was right, that his theory was as good as Schwinger's, but somehow he couldn't get the message across. He decided to go back to Cornell and write it all up for publication, so that they could study it in cold print.[21]

But the Pocono Conference was far from being a disaster for Feynman. In the intervals between formal lectures, over lunch and coffee and whenever they could get together, he and Schwinger compared notes. Neither of them really understood what the other was doing, but they trusted and respected each other. For every problem that they had both tackled, it turned out that they had got the same answer:

> We came at things entirely differently, but we came to the same end. So there was no problem with my believing that I was right and everything was OK.[22]

To Feynman and Schwinger, being told the same thing twice by the equations meant it must be true. In Lewis Carroll's *The Hunting of the Snark*, 'what I tell you three times is true'. The third telling of the story of QED was about to happen in spectacular fashion.

Oppenheimer was by now Director of the Institute for Advanced Study, and when he got back to Princeton after the Pocono Conference he found a letter and a package of scientific papers waiting for him. They came from a Japanese physicist, Sin'Itiro Tomonaga, who had worked out essentially the same version of QED as Schwinger, largely cut off from contact with Western scientists, in the harsh conditions of battered wartime and postwar Tokyo. This incredible achievement has been described in detail by Silvan Schweber, in *QED and the Men Who Made It.* Tomonaga had not only come up with a slightly simpler version of QED than Schwinger (proof, if any were needed, of Schwinger's love of sometimes unnecessary complications) but he had actually been the first of the three physicists to complete his theory.

The physics community had been told three times that QED was true, and it was. But how did Feynman's version of QED come to be recognized, before long, as the simplest approach, a break with tradition that, instead of being the last flowering of an old glory, provided a seed from which great new ideas would grow? Feynman did indeed start to publish his work in a series of clear and impressive papers after the Pocono Conference. But the key to getting his message across to a wider audience owed much to the presence in Princeton of another mathematical prodigy, the Englishman Freeman Dyson. Where Schwinger had demonstrated his talent by finishing his PhD work before completing his BSc, Dyson would demonstrate his in equally impressive fashion by (eventually) becoming a member of the Institute for Advanced Study without finishing a PhD at all.

Dyson had been born in 1923, and after graduating from

Cambridge worked for the British wartime Bomber Command on statistical studies of the effectiveness of the bombing campaign over Germany. This was a doubly futile exercise – it was a waste of Dyson's mathematical talent, and he soon discovered (although he was never able to convince his superiors) that the bombing effort was largely misdirected and a waste of the lives of inexperienced aircrew sent on impossible missions. In September 1947, he enrolled as a graduate student in the physics department at Cornell, working under Bethe and in an ideal position to observe the dramatic development of QED over the next few months. He has often told the story, most notably in his book *Disturbing the Universe*,[23] from which the following account is largely drawn.

The first task Bethe gave Dyson was to redo Bethe's calculation of the Lamb shift for a spin-zero electron (a fictitious simplification), with the requirements of the Special Theory of Relativity (corresponding to taking note of the spin) bolted on in an ad hoc fashion. This provided no new insight into the quantum world, but after hundreds of pages of calculations Dyson ended up with what he describes as a 'pastiche', no real improvement on Bethe's calculation, which more or less gave the right answer. A good analogy with Bethe's and Dyson's efforts to explain the Lamb shift would be the Bohr model of the atom, a patchwork of ideas put together on an ad hoc basis which worked after a fashion, but gave no deep insight into what was going on. The hours Dyson spent in this calculation were, though, a valuable familiarization with the cutting edge of what was going on in quantum physics at the time. Dyson was too junior a researcher to be present at the Pocono Conference, but he was well aware of Feynman as 'the liveliest person in our department', who 'refused to take anybody's word for anything' and had set out 'to reinvent quantum mechanics'.

Dyson soon realized that Feynman, with his new quantum mechanics, could solve every problem that Bethe could solve using the older version, getting the same answers. But Feynman could also solve a lot of problems that the old quantum mechanics couldn't handle. 'It was obvious to me that Dick's theory must be fundamentally right. I decided that my main job, after I finished the calculation for Hans, must be to understand Dick and explain his ideas in a language that the rest of the world could understand.'

It seemed that Dyson might not get the opportunity to do this, because after a year at Cornell he was scheduled to spend a year doing research at the Institute for Advanced Study, working with Oppenheimer. This left him only a few months to try to get to grips with Feynman's work. He made an effort to see Feynman as much as possible, and happily accepted the way in which Feynman dealt with visitors. If he didn't want to be disturbed, he would just shout, 'Go away; I'm busy.' But if he let you into his office, it meant that he really did have time to talk. They talked for hours about Feynman's theory, until Dyson began to feel that he was beginning (only beginning) to get the hang of it – but his time in Cornell was nearly up.

The reason why ordinary physicists had trouble getting hold of Feynman's ideas, Dyson realized, was that Feynman thought in pictorial terms. He had a physical picture of how the world worked, a picture which gave him an insight into the solution of complicated problems without having to write down a lot of equations. In an interview with Silvan Schweber,[24] Feynman said:

> Visualization in some form or other is a vital part of my thinking . . . half-assed kind of vague, mixed with symbols. It is very difficult to explain, because it is not clear. My atom, for example, when I think of an electron spin in an atom, I see an atom and I see a vector and a ψ written somewhere, sort of, or mixed with it somehow, and an amplitude all mixed up with *x*s . . . it is very visual . . . a mixture of a mathematical expression wrapped into and around, in a vague way, around the object. So I see all the time visual things associated with what I am trying to do.

In *What Do You Care*, Feynman tried once again to explain how he thought abut physics:

> When I see equations, I see the letters in colors – I don't know why. As I'm talking, I see vague pictures of Bessel functions from Jahnke and Emde's book, with light-tan *j*'s, slightly violet-bluish *n*'s, and dark brown *x*'s flying around. And I wonder what the hell it must look like to the students.

Another great physicist who also thought in visual terms was Albert Einstein, although his pictures – a person riding on a beam of light, or falling in an elevator with a broken cable – seem to have been more clear-cut and down to earth than Feynman's.

The term at Cornell ended in June, and Dyson still hadn't got Feynman's new quantum theory straight. Thanks to Bethe, he had an opportunity to attend the summer school at the University of Michigan in Ann Arbor, the latest in a series of gatherings famous since 1930, where Schwinger would be giving a full account of his version of QED. He had two weeks to kill before the summer school started, and when Feynman invited Dyson to join him on a drive over to New Mexico, Dyson leapt at the chance.

The reason for the trip back to Albuquerque was a girl, a young woman that Feynman had dated after the death of Arline, and with whom, for a time, he imagined that he might settle down. In a letter home to his parents in England,[25] Dyson mentioned the difficulties involved. 'The girl is a Catholic. You can imagine all the troubles this raises, and if there is one thing Feynman could not do to save his soul it is to become a Catholic himself.'

As far as love was concerned, the trip was a waste of time. Feynman and the girl no longer felt the old attraction for one another, and the question of marriage never seriously arose. But on the way to Albuquerque Dyson had Feynman to himself (along with the occasional hitchhiker they picked up) for four whole days, discussing life and physics. In the middle of Oklahoma, they ran into torrential rain and flooding so bad that all progress was halted, and ended up in a place called Vinita looking for a room for the night. The town was packed with other stranded travellers, and all the hotels were full. Feynman, though, was unfazed. From his days seeking out the cheapest possible accommodation for his overnight visits to be near Arline, he knew what to do, and found them a room in a brothel that they could share for 50 cents apiece.

With the rain hammering down and the girls plying their trade in nearby rooms, there was no prospect of sleep that night, but the two travellers were happy simply to be warm and dry. They talked the night away – or rather, Feynman talked and Dyson mostly listened. He talked about Arline, and his work on the bomb. Then they talked about physics, and Dick's way of visualizing quantum processes in spacetime. Dyson saw that Feynman's sum over histories theory 'was in the spirit of the young Einstein'. But 'nobody but Dick could use his theory, because he was always invoking his intuition to make up the rules of the game as he went along. Until the rules were codified and made mathematically precise, I could not call it a theory.'[26]

The next day the rains had eased and the roads were passable. In Albuquerque, they made their farewells, and Dyson caught the Greyhound bus back east, travelling in easy stages to Ann Arbor and revelling in his first experiences of travelling alone across America. In five weeks at Ann Arbor, as well as attending lectures he made many new friends, and managed to talk at length with Schwinger about his theory. At the end, 'I understood Schwinger's theory as well as anybody could understand it, with the possible exception of Schwinger.' From Ann Arbor, Dyson travelled back across the United States by Greyhound to holiday in San Francisco. At the beginning of September, it was time to head east once again, to Princeton. For three days and nights he travelled non-stop, as far as Chicago. He had nobody to talk to, the roads were too bumpy for sleep, and he:

> looked out of the window and gradually fell into a comfortable stupor. As we were droning across Nebraska on the third day, something suddenly happened. For two weeks I had not thought about physics, and now it came bursting into my consciousness like an explosion. Feynman's pictures and Schwinger's equations began sorting themselves out in my head with a clarity they had never had before. For the first time I was able to put them all together. For an hour or two I arranged and rearranged the pieces. Then I knew that they all fitted. I had no pencil or paper, but everything was so clear I did not need to write it down. Feynman and Schwinger were just looking at the same set of ideas from two different sides.

So it was that Dyson wrote up a paper on 'The Radiation Theories of Tomonaga, Schwinger and Feynman', and sent it off to the *Physical Review* even before Oppenheimer returned from his own summer travels in Europe. The paper[27] made the new quantum electrodynamics, at last, accessible to ordinary physicists, and made Dyson's reputation, although Oppenheimer, as it turned out, needed a lot of convincing that it was all worthwhile. By now, Feynman was also pressing ahead with writing up his work for publication, and had sorted his ideas into a much more clear and accessible form than the messy failure of a lecture that he gave at the Pocono Conference.

Because of Dyson's comprehensive and influential review of the whole field, though, some people were initially confused about who had discovered (or invented) what, and for a time what are now known as 'Feynman diagrams' (which we discuss in the next chapter) were referred to in some quarters as 'Dyson graphs'. It didn't matter. Feynman and Schwinger were both happy to see their work receiving the attention that it deserved. As Steven Weinberg has observed, 'with the publication of Dyson's papers, there was at last a general and systematic formalism that physicists could easily learn to use, and that would provide a common language for the subsequent applications of quantum field theory to the problems of physics'.[28] Or as Dyson himself has put it, 'my major contribution [was] to translate Feynman back into language that other people could understand . . . When Feynman's tools first became available, it was a tremendous liberation – you could do all kinds of things with them you couldn't have done before.'[29]

Almost immediately, Dyson was given a demonstration of the power of Feynman's toolkit when wielded by Feynman himself. At the end of October, when Dyson had finished his paper, he visited Cornell with another physicist from the Institute, Cecile Morette, to discuss quantum electrodynamics and make sure there were no hard feelings about what he had done – that he had written an account of Feynman's theory before Feynman had published it himself. Dyson had sent Feynman a copy of the paper, which Feynman had given to one of his students to read. He then asked the student if there was any need to read it himself, and the student said no, so he didn't.[30] Dyson and Morette arrived at Cornell on a Friday, and were entertained by Feynman with stories and drumming until 1 a.m. The next day, he gave them a 'masterly account' of his theory. In the evening, Dyson mentioned that there were two outstanding problems that the theory had yet to tackle, and which had proved intractable for the old theories in spite of intensive efforts by many physicists. They were problems involving the scattering of light (photons) by an electric field, and the scattering of photons by other photons. 'Feynman said "We'll see about this," and proceeded to sit down and in two hours, before our eyes, obtain finite and sensible answers to both problems. It was the most amazing piece of lightning calculation I have ever witnessed', Dyson wrote to his parents, 'and the results prove, apart from some unforeseen complications, the consistency of the whole theory.' Years later, in a TV interview,[31] Dyson described this as 'just about the most dazzling display of Feynman's powers I've ever seen. These were problems that had taken the greatest physicists months to fail to solve, and he knocked them off in a couple of hours . . . it was done in this extraordinary economical style, without heavy apparatus – just sort of stitching the answers before even writing down the equations, and deriving things directly from the diagrams. Well, after that there was nothing more to be done, but only to proclaim the triumph of the theory.'

This was Feynman at the height of his powers, delighting in applying his new theory to solving problems. This particular feat impressed Dyson; but Feynman managed to impress himself with his next *tour de force*, which took place at the January 1949 meeting of the American Physical Society. At the meeting, a physicist named Murray Slotnick presented some new results describing the way an electron bounces off a neutron. He had calculated these the old way, over a period of many months. Feynman missed the talk, but was told about it by a colleague. He asked Slotnick how he had approached the problem, and decided it would be 'a welcome opportunity' to test his theory by seeing if it gave the same answers. In his Nobel lecture, Feynman describes how he worked through the problem that evening, and next day went up to Slotnick to compare notes. Slotnick said, 'What do you mean, you worked it out last night, it took me six months!' And when they checked, they found that not only had Feynman got the same answers as Slotnick, but that he had solved the problem in a much more general way, allowing for the momentum transferred by the electron to the neutron (the recoil of the neutron when hit by the electron); Slotnick had only solved the problem for zero momentum transfer (no recoil).

This, Feynman recalled in his Nobel lecture, was the moment when everything came together for him. 'That convinced me, at last, that I did have some kind of method and technique and understood how to do something that other people did not know how to do. That was my moment of triumph.'

The work was published in a series of papers over the next three years, but by the beginning of 1949 everything was complete. Neatly rounding off this epic period in the development of quantum theory, the third, and last, of the postwar conferences organized by Oppenheimer and funded by the National Academy of Sciences took place from 11 to 14 April 1949, at Oldstone-on-the-Hudson in Peekskill, New York, 50 miles north of New York City. By now, Dyson was eminent enough to be included in the couple of dozen participants, and whereas Schwinger's theory had formed the centrepiece of the Pocono Conference, while the Lamb shift had been the main talking point at Shelter Island, at the Oldstone Conference it was Feynman's approach to QED that was at centre stage. A month before his thirty-first birthday, Feynman had become the leading physicist of his generation, pointing the way ahead with his new ideas.

Shortly after the Oldstone gathering, Dyson gave a talk in Washington to a meeting of the American Physical Society, at which he said:

> We have the key to the Universe. Quantum electrodynamics works and does everything you wanted it to do. We understand how to calculate everything concerned with electrons and photons. Now all that remains is merely to apply the same [ideas] to understand weak interactions, to understand gravitation and to understand nuclear forces.[32]

These seemingly extravagant claims have largely been proved correct; although gravity has not yielded to the attack as easily as Dyson hoped in 1949, all of the rest of physics is now understood in the same terms as Feynman's formulation of QED. Before we look at how Feynman's life and career developed after 1949, it is worth taking stock of the breathtaking way in which QED, and especially Feynman's formulation of QED, has played the central role in all of theoretical physics (except the investigation of gravity) throughout the second half of the twentieth century.

Notes

1. Reprinted in *Surely You're Joking*, from *Reminiscences of Los Alamos, 1943–45*, edited by Lawrence Badash, Joseph Hirschfelder and Herbert Broida (Reidel, Dordrecht, 1980). Unless otherwise indicated,

anecdotes about Feynman's time in Los Alamos come from this source.

2. *Scientific American*, June 1988.
3. Personal correspondence between Arline and Richard, transcribed and loaned by Michelle Feynman.
4. See Bethe's contribution to *Most of the Good Stuff.*
5. See note 4.
6. *What Do You Care.*
7. *What Do You Care.*
8. See note 3.
9. *What Do You Care.*
10. See note 1.
11. *Robert Oppenheimer: Letters and Recollections*, edited by Alice Kimball Smith & Charles Weiner (Harvard University Press, 1980).
12. Mehra.
13. Interview with Silvan Schweber, reported in *QED and the Men Who Made It.*
14. See note 3.
15. *Surely You're Joking.*
16. Nobel lecture.
17. Nobel lecture.
18. Mehra.
19. Quoted by Mehra.
20. Mehra.
21. Mehra.
22. Feynman, quoted by Mehra.
23. Freeman Dyson, *Disturbing the Universe* (Basic Books, New York, 1979).
24. See *QED and the Men Who Made It.*
25. Reprinted in Freeman Dyson, *From Eros to Gaia* (Pantheon, New York, 1992).
26. See note 23.
27. *Physical Review*, volume 75, page 486, 1949.
28. Steven Weinberg, *The Quantum Theory of Fields* (Cambridge University Press, 1995).
29. See Dyson's contribution to *No Ordinary Genius.*
30. See note 25.
31. See *No Ordinary Genius.*
32. Quoted by Schweber.

— 6 —
The masterwork

Quantum electrodynamics is a theory that describes all interactions involving light (photons) and charged particles, and, in particular, all interactions involving photons and electrons. Because the interactions between atoms depend on the arrangement of electrons in the clouds around the nuclei, that means that, among other things, QED underpins all of chemistry. It explains how a spring stretches, and how dynamite explodes; the working of your eye, and how grass is green (it is also the explanation behind the intermolecular forces described in Feynman's undergraduate thesis). In fact, as far as the everyday world is concerned, QED explains everything that isn't explained by gravity. There are two other forces of nature, which only operate on a very small scale, essentially within the nucleus of an atom, and are responsible for holding those nuclei together and for radioactivity. But outside the nucleus, on the scale of atoms and above, all that matters is QED and gravity.

Both QED and gravity (in the form of Einstein's General Theory of Relativity) are extremely accurate and well-understood theories. In terms of experiments actually carried out in laboratories here on Earth, though, QED is the outstanding example of a successful theory – that is, one which predicts with great precision the outcome of experiments. The property called the magnetic moment of the electron, which we mentioned in Chapter 5, is, along with the Lamb shift, a classic example of how the new theory achieved such success, and one which can be explained neatly using Feynman's techniques. Using Dirac's theory of the electron, you can choose to work in units in which the value of the magnetic moment of the electron is precisely 1. QED, however, predicts a value of 1.00115965246, while experiments have measured the magnetic moment to be 1.00115965221. The uncertainty in the experimental measurement is about ±4 in the last number; the uncertainty in the theoretical calculation is about ±20 in the last two numbers. So theory and experiment agree to an accuracy of two parts in 10 decimal places, or 0.00000002 per cent. In his book *QED: The Strange Theory of Light and Matter,*[1] Feynman points out that this is equivalent to measuring the distance from Los Angeles to New York to the thickness of a human hair – and this is just one example of the many precise agreements between QED and experiment. Very recently, the General Theory of Relativity has been checked to a similar accuracy by studying the behaviour of an astronomical object known as the binary pulsar; but somehow, that isn't quite the same as doing the experiments, for real, right here on Earth. In that sense, QED is the most successful and accurate of all scientific theories, although both kinds of observation are really equally valid.

Feynman's approach to QED, using path integrals, can best be seen by starting out with the famous experiment with two holes, discussed in Chapter 2. The important thing about the experiment with two holes – thinking for the moment in terms of waves – is that waves that follow one path through the experiment to the detector screen can get out of step with waves that follow the other path through the experiment. Waves that march in step with one another are said to be in phase, and if both waves are of equal strength *and* in phase, they will add together to produce a wave that is twice as strong. But if two waves of equal size have opposite phase (that is, they are exactly out of phase), then they will cancel out. It is this addition and cancelling of waves that makes for the pattern of bright and dark bands on the screen in the experiment with two holes, even though all the waves have the same strength. It is also the difference in phase, not any difference in the strength of the waves, that makes the advanced and retarded waves in the absorber theory of radiation add up and cancel out in just the right way to explain how charged particles interact (see Figure 5). And, of course, as well as complete addition and cancellation it is possible to have intermediate cases, where two waves are out of phase but not perfectly opposed to one another, producing a partial cancellation.

All of this carries over to the alternative quantum mechanical description of what is going on, where the light is described in terms of entities (photons, electrons or anything else) that follow trajectories determined by quantum probabilities. These quantum probabilities are described by the Schrödinger equation, and behave exactly like waves, with phase all important in determining whether

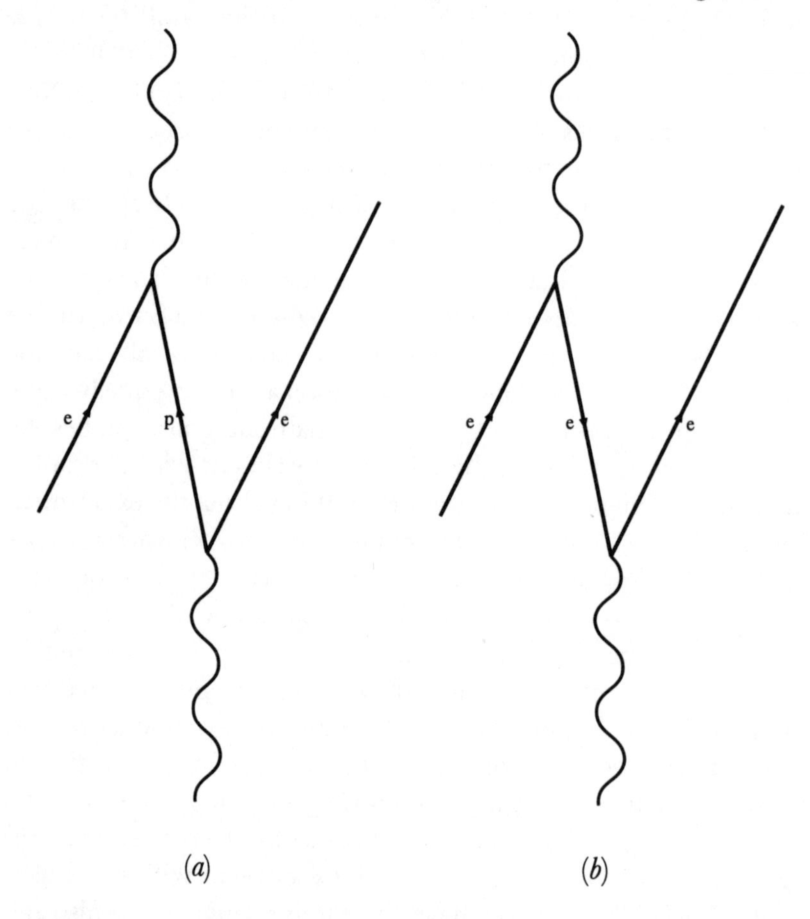

Figure 7. (*a*) A spacetime diagram can be used to show how an energetic photon (a gamma ray) can give up its energy to create an electron (e) and a positron (p). The positron later meets another electron and annihilates with it, to make a gamma ray. (*b*) But it is equally valid to say that there is only one electron (e), which starts out on the left moving into the future, then meets an energetic photon travelling backwards in time, which sends the electron moving backwards in time until it meets another gamma ray which bounces it back to the future. A positron is an electron going backwards in time.

two probabilities add up to produce a strong likelihood of a photon (or whatever) following a particular path, or cancelling out to ensure that the photon never takes a second path. The only slight complication is that the actual probabilities are given in terms of the square of the wave property known as the amplitude – the probability amplitudes have to be combined first (putting the little arrows head to tail), and the answer you get then multiplied by itself to give the actual probability of a particular path being followed.

The experiment with two holes shows that, even for an entity we are used to thinking of as a particle (such as an electron), something (either the particle itself or the probability wave) goes through both holes in the experiment and interferes with itself in this way to determine the pattern on the screen. But suppose we make an experiment with four holes, instead of just two. Now, obviously, the 'something' has to travel through all four holes and make the appropriate interference pattern, and this can be calculated using the rules we have just sketched out. The same is true for an experiment with three holes, or a hundred, or any number you like. You can imagine making more and more holes until there is nothing left to obstruct the path of the electrons or photons at all – you have an experiment with no holes, or one hole, or infinitely many holes, depending on your point of view. One of Feynman's key insights was that you can still treat the electron or photon (or anything else) as having gone through each of the infinite number of holes, adding up the probabilities associated with each path in the usual way. Integrating (adding up) the probabilities for literally every possible path from the source of the light or electrons to the detector screen on the far side of the experiment then gives you the result that the overwhelmingly most probable path for the particle to follow is a straight line from the source to the detector. For more complicated paths, the phases of adjacent trajectories are exactly opposed to one another (the arrows point in opposite directions), and they all cancel out, leaving just the path that is expected from classical physics. It is only near the classical path (the path of least action) that the probabilities add up and reinforce one another, because they are in phase. And so Feynman's path integral approach to quantum mechanics does indeed also give classical mechanics, and all of classical optics, from the same set of equations.

This is such a dramatic discovery that it is worth showing one example of how it makes us think again about familiar features of the world, such as the idea that 'light travels in straight lines'. In Figure 8, we show how classical optics teaches us that light is

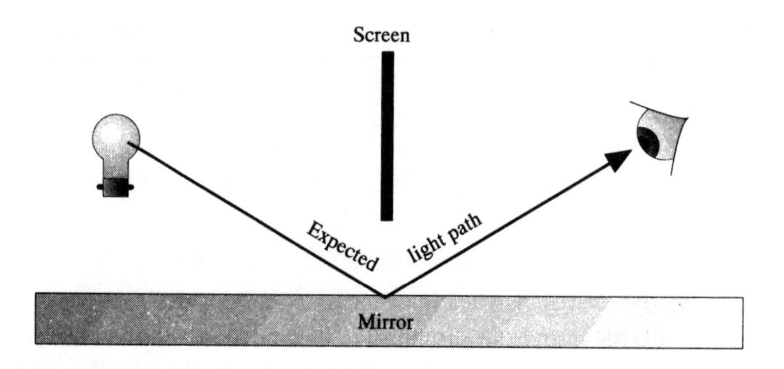

Figure 8. Common sense (and schoolbook physics) tells us that 'light travels in straight lines'.

reflected from a mirror. This is so familiar that it seems to fly in the face of common sense to suggest that the image you see in the mirror is a result of light coming from the source in all directions, bouncing off the mirror at all kinds of crazy angles and reaching your eye that way, as it looks in Figure 9. That, though, is exactly what happens, according to Feynman. But the light travelling by crazy angles gets cancelled out by neighbouring light that is equally

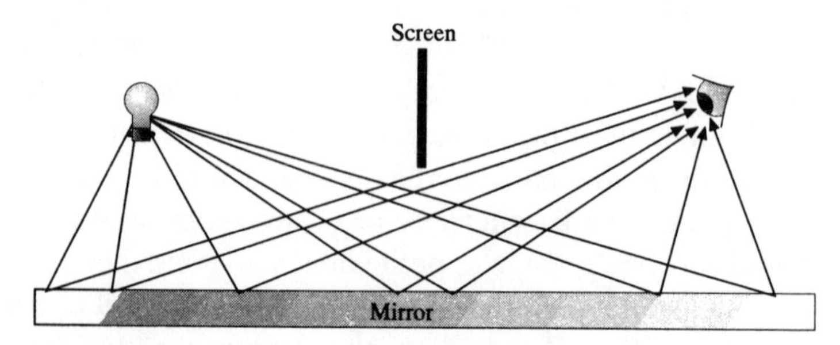

Figure 9. Feynman says that light travels by every conceivable crazy path from the source to your eye, bouncing off the mirror at all kinds of angles (and even travelling by weird routes that do not involve bouncing off the mirror at all).

strong but has opposite phase, so that you do not become aware of it. Because of phase differences, the amplitudes only add up and reinforce each other near the path of least time from the source to your eye – the Principle of Least Action is at work, and as Feynman put it in *QED*, 'where the time is least is also where the time for nearby paths is nearly the same', which is why the probabilities add up there.

You can actually prove, by yourself, that light from the edges of the mirror really is entering your eyes by some of the crazy routes shown in Figure 9. In the more scientifically precise version of such an experiment, you first cover up all the mirror except for a bit out by the edge, so that it cannot reflect. Way out on the edge of the mirror, although the probabilities for neighbouring paths cancel out, you can still find thin strips of mirror where the probabilities all add up. The trouble is, these strips are separated from one another by equally thin strips for which the probabilities are exactly out of phase with the first set of strips, so you see no light from the edge of the mirror. All you have to do, though, is cover up alternating strips of mirror. You are left with half as much working mirror, but now all the paths are in phase, and you really will actually see the light coming to you from these crazy angles (Figure 10).

The set-up is called a diffraction grating, and because the effect depends to some extent on the wavelength of the light, if you do it with ordinary light you will see a colourful rainbow pattern. And you don't even have to go to the trouble of laying out a mirror and

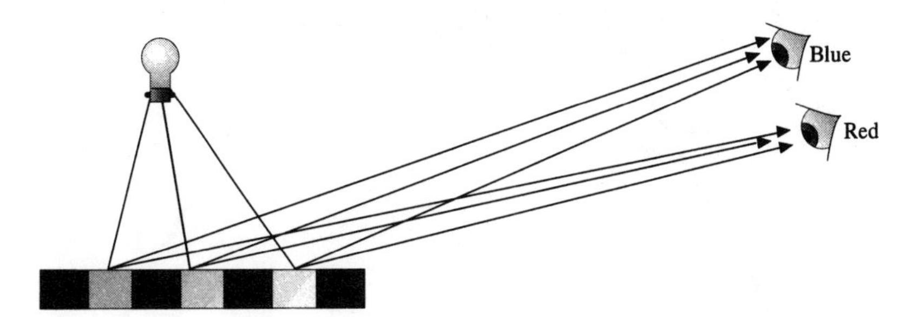

Figure 10. We don't normally see light bouncing off mirrors at crazy angles because the light cancels out everywhere except near the path of least time. But if strips of mirror are carefully blacked out to stop the cancelling, light really is seen to be reflected at all kinds of weird angles.

covering it with strips of cloth carefully cut to a precise width. The spacing you need to produce the effect with ordinary light is the same as the spacing of the grooves on an ordinary compact disc. Just hold a CD under the light, and you will see for yourself a rainbow pattern caused by photons bouncing off the disc at the 'wrong' angles – quantum electrodynamics made visible in your own home. Whether taking the path of least time or bouncing around at 'crazy' angles, 'Light doesn't *really* travel only in a straight line', said Feynman; 'it "smells" the neighbouring paths around it, and uses a small core of nearby space.'

Which brings us on to the famous Feynman diagrams. The archetypal Feynman diagram is a spacetime diagram which represents an interaction between two electrons that involves the exchange of a photon. The electrons approach one another, exchange the photon, and move apart (Figure 11). But there is much more to this kind of diagram than appears at first sight. For a start, the exchange of the

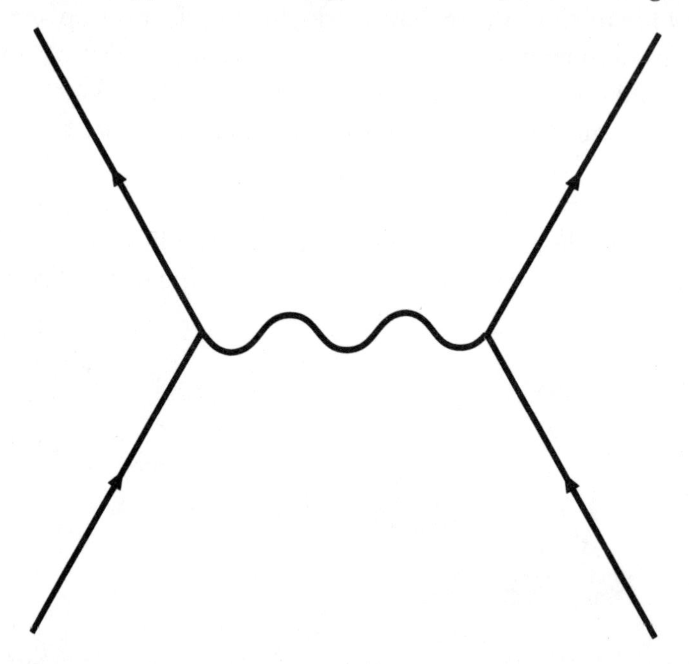

Figure 11. The archetypal Feynman diagram. Two particles (perhaps two electrons) approach one another, interact by the exchange of a force-carrying particle (in this case, a photon) and are deflected.

photon represented by the wiggly line should not be taken as a 'classical' particle following a single spacetime path, but as the sum over histories of all possible ways in which that photon could have gone from one particle to the other. The wiggly line doesn't represent a path, but a summation of all possible paths – a path integral. Secondly, what goes on at the junctions of a Feynman diagram, where different lines intersect, is precisely determined by the rules of quantum electrodynamics. Each kind of intersection – each vertex – represents a different kind of interaction, each with its own precise meaning and its own set of equations that describe what is going on. In this sense, a few Feynman diagrams can represent a kind of shorthand for the hundreds of equations required by Schwinger's or Tomonaga's approach to QED. In January 1988, Feynman stressed that:

> The diagrams were intended to represent physical processes *and the mathematical expressions* [our italics] used to describe them. Each diagram signified a mathematical expression. Mathematical quantities were associated with points in space and time. I would see electrons going along, being scattered at one point, then going over to another point and being scattered there, emitting a photon and the photon goes over there. I would make little pictures of all that was going on; these were physical pictures involving the mathematical terms. These pictures evolved only gradually in my mind . . . they became a shorthand for the processes I was trying to describe physically and mathematically . . . I was conscious of the thought that it would be amusing to see these funny-looking pictures in the *Physical Review*.[2]

One of the most important features of these diagrams is that they treat particles and antiparticles on an equal footing, which is what makes Feynman's theory Lorentz invariant, in line with the requirements of relativity theory. By treating particles and antiparticles in the same way, the nature of the infinities that arise in QED becomes clear (at least to a mathematician), and Freeman Dyson proved that the infinities that arise in interactions described by Feynman diagrams are always of the kind which can be removed by renormalization – a dramatic result which did much to persuade other physicists of the value of Feynman's approach. Today, one of the chief criteria used to decide whether a new idea in particle physics is worth pursuing is whether or not the theory is renormalizable – that is, whether or not it can be described using Feynman diagrams. If it cannot, then it is rejected out of hand.

Feynman's 'funny-looking pictures' have become so important both because they really do incorporate all of the complex mathematical rules, and because they give a direct practical insight into what is going on. To use them properly (to get numbers out of the calculations to compare with experiments), you need to understand the mathematics. But to get an idea of what is going on, you only need the pictures – and that's all we are going to be concerned with now as we indicate how that fantastically accurate calculation of the magnetic moment of the electron was worked out. With the physical insight provided by the pictures, Feynman diagrams can even give a picture of processes too complicated to be calculated, but which have a clear physical meaning that could only be derived from Schwinger's pages of equations by a virtuoso mathematician. To a virtuoso, this democratization of physics may seem unnecessary; many years later, Schwinger described the effect of the Feynman diagram as 'bringing computation to the masses';[3] he did not intend this as a compliment.

The simplest version of the interaction between an electron and the field of a magnet can be represented in a diagram like Figure 12.

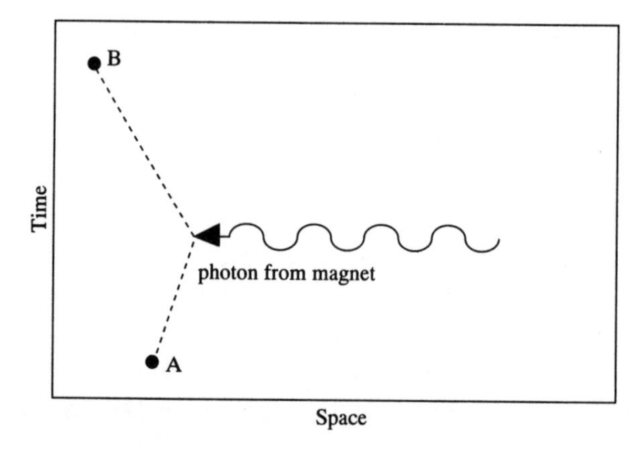

Figure 12. A Feynman diagram can also describe how an electron moving from A to B is deflected when it interacts with a magnetic field (when it meets a photon from a magnet).

A photon from the magnet is absorbed by the electron. If the situation were really that simple, the calculated magnetic moment of the electron would be 1. In fact, as we have mentioned, it is actually a little bigger, about 1.00116. But the electron can also be involved in a kind of self-interaction, in which it emits a photon and later reabsorbs the same photon (called a 'virtual' photon), while in between, it interacts with the photon from the magnet. This is represented in a Feynman diagram like Figure 13. And when you

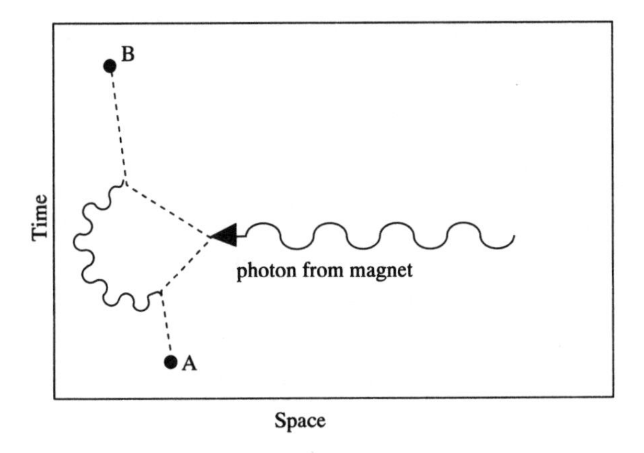

Figure 13. Things are not quite as simple as they seem in Figure 12. The electron can emit a virtual photon, and then reabsorb it, as well as interacting with the photon from the magnet. More and more complicated loops can be added, but happily in this case they have smaller and smaller influences on the interaction.

do the corresponding calculation, you get a value for the magnetic moment, allowing for all possible interactions of this kind, a bit bigger than 1, but still not quite as big as the experimental value. It was this single virtual photon version of the calculation that showed physicists they were on the right track in the 1940s.

Of course, the next step in the process is obvious. You have to consider the possibility that the electron emits two photons, one after the other, and reabsorbs them. Sure enough, when you do the calculation you get an answer a little closer to the experimental figure. But now the calculations are getting difficult, and it took two years for all the possibilities involving two of these virtual photons to be included. It wasn't until the middle of the 1980s that the calculation involving up to three virtual photons was carried through, giving the value for the magnetic moment that we quoted at the beginning of this chapter, in very close agreement with the experiments. And, equally significantly, we can see immediately why the theory doesn't yet give precise agreement with experiment – we have not yet included the effects of four virtual photons, or five, or still greater numbers. Happily though, the correction gets smaller for each extra photon in the calculation, and the results for three virtual photons are good enough to satisfy most people.

It's just as well that the correction gets smaller for higher numbers of virtual photons – for higher 'order' in the calculation – because there are yet further complications that really ought to be included, if not in the calculations then at least in our mental picture of what is going on around an electron, or any other quantum entity. It's easy to think that you understand where the energy required to make a virtual photon can come from. A single photon doesn't carry a lot of energy, and no doubt the electron can spare some of its kinetic energy, or whatever, to make the photon. But this isn't quite the right picture.

There is one key ingredient of quantum mechanics that we have not yet discussed, and it is called uncertainty. In the quantum world, it turns out, it is impossible for all of the properties of a quantum entity, such as a photon or an electron, to be specified at the same time. This restriction was first worked out, in the 1920s, by Werner Heisenberg, and is known as Heisenberg's Uncertainty Principle, or just as the Uncertainty Principle. The important point is that it has nothing to do with our clumsiness in trying to make measurements of the properties of tiny things like electrons; it is built into their very nature.[4] So, for example, an electron cannot have both a precise location in space and a precise momentum (a definite direction) *at the same time*. It may have a very well-defined location (as when it makes a spot of light on a detector screen), but then *the electron itself* cannot 'tell' where it is going next. Or it may have very well-defined momentum, as when it is travelling along a certain trajectory, but then *the electron itself* does not 'know' exactly where it is along that trajectory.

Uncertainty also applies to the energy available to make virtual particles. According to the Special Theory of Relativity, you need a certain amount of energy, mc^2, to make an electron. In fact, since the quantum rules only allow the creation of electron–positron pairs, you need $2mc^2$ to make the pair. But quantum uncertainty says that for a short enough time (a *very* short time!) the Universe cannot be certain that there isn't that much energy in any tiny volume of empty space. So electron–positron pairs can be created anywhere and everywhere, *provided that they almost immediately get back together and annihilate one another*. The more energy you 'borrow', the quicker you have to pay it back.

This is where virtual photons actually 'come from'. They don't have to borrow any energy from the electrons involved in an interaction. They borrow it from empty space – from nothing at all – while, in a sense, the Universe isn't looking. Because photons carry little energy, virtual photons can be made in profusion in this way, and last for a relatively long time. But quantum uncertainty says that during its existence, the low-energy photon can, very briefly, borrow a lot more energy from nothing at all, and turn itself into an electron–positron pair. The pair promptly gives back the energy and disappears, turning back into a photon, but the process can repeat during the lifetime of the virtual photon. And even these virtual electrons and virtual positrons can be involved in the whole business of creating photons and virtual pairs. Each 'real' electron is actually surrounded by a frothing cloud of virtual photons and other entities, popping in and out of existence all the time.

In spite of this complexity, QED is so good that it can be used to calculate, with the aid of Feynman diagrams, all kinds of messy interactions involving photons being exchanged between charged particles. It is the cloud of virtual photons (and other things) around an electron which prevents it from behaving as a 'bare' point charge and reduces the self-interaction from infinity to a small amount responsible for the Lamb shift. But QED can do more than explain everything there is to explain about the behaviour of photons and electrons. It provides the template with which physicists have built their theories of the workings of those other forces we mentioned, the ones that operate within the nucleus.

One of these forces is called the strong interaction, because it is the strongest of all the four forces of nature. It is an attractive force that holds the nucleus together, operating on both neutrons and protons and overcoming the electrical repulsion between all the positively charged protons in the nucleus, which tries to blow the nucleus apart. The other nuclear force is called the weak interaction, because it is weaker than the strong interaction. Very little was known about the weak interaction in the 1940s, but after the success of QED in explaining electromagnetism, in the 1950s many physicists worked on the problem of developing a deeper understanding of the force – Feynman was also involved in some of this work, as we shall see in Chapter 8. Two physicists, Abdus Salam and Steven Weinberg, independently cracked the problem in the 1960s, and shared the Nobel Prize for their efforts in 1979. Again, we won't go into the (sometimes hairy) mathematical details; the relevant point is that the resulting theory of the weak interaction is exactly like the QED theory of electromagnetism, and can be understood in terms of Feynman diagrams involving a greater variety of particles (which is one reason why the mathematics is hairy).

The particles that can take part in weak interactions are the proton and neutron, on one side, and the electron and an associated particle called the neutrino on the other side. Protons and neutrons are members of a family called baryons, and electrons and neutrinos are members of a family called leptons. Moving between the two families there are so-called intermediate vector bosons, which play the role in the weak interaction that photons do in electromagnetism – only there are *three* kinds of vector boson, one with zero charge (dubbed Z^0), one carrying a unit of positive charge (dubbed W^+), and one carrying a unit of negative charge (the W^- boson). Unlike photons, these bosons each have mass. There is one other important rule. The total number of baryons involved in an interaction always stays the same, and the total number of leptons always stays the same.

The basic process of radioactive decay is seen at its most simple when a neutron sits on its own, outside an atom. Within a few minutes, the neutron will decay, spitting out an electron and transforming itself into a proton. Electric charge is conserved, because the positive charge on the proton and the negative charge on the electron cancel out. The number of baryons is conserved, because you start with one (a neutron) and end up with one (a proton). At first sight, it seems that the world has gained a lepton (the electron); but it turns out that in neutron decay another particle, an antineutrino, is always produced as well. So there are still zero leptons overall, since a particle and an antiparticle cancel each other out, for these purposes, in the same way that the positive charge and the negative charge cancel each other out.

In order to represent this on a Feynman diagram, you can use one of Feynman's neat tricks. An antiparticle leaving the neutron and heading into the future is the same as a particle arriving at the neutron from the past. In Feynman's world, the prefix 'anti' on a particle's name means 'going backwards in time'. So the fundamental example of the weak interaction at work is represented by a diagram like Figure 14. The key point is that this description of

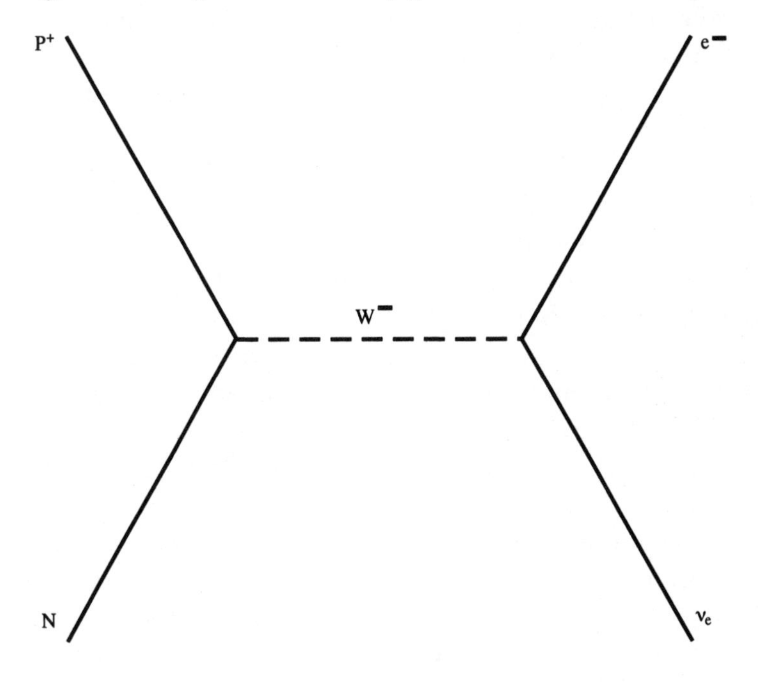

Figure 14. Using QED as its template, the electroweak theory describes an interaction in which a neutron (N) interacts with a neutrino (ν_e) by the exchange of a W^- particle to produce a proton (p^+) and an electron (e^-) (compare with Figure 11). Such a Feynman diagram can be read equally validly 'down the page', with (in this case) an electron and a proton interacting to produce a neutron and a neutrino.

the weak interaction is *exactly* the same as QED, once allowance is made for the extra particles and their properties. It even includes the same kind of infinities as QED, which are removed in the same way, by renormalization. Among other things, this means that all the arrows on the diagram (on *any* Feynman diagram) can be reversed to describe an equally valid fundamental interaction – in this case, a proton and an electron can interact, with the exchange of a W^- particle, to make a neutron and a neutrino.

The match between the rules of the weak interaction and the rules of QED is so exact, in fact, that there is no point in trying to pretend that they are different theories. Today, physicists speak of the 'electroweak' theory of particle physics, one set of equations that describes all interactions involving either electromagnetism or the weak interaction, or both (including, remember, all of classical mechanics, in Feynman's formulation of QED). That set of equations (and diagrams) is essentially the QED template itself. As well as explaining everything there is to explain about interactions involving electrons and photons, the QED template explains everything there is to explain about weak interactions, to almost the same high precision as QED itself.

The situation isn't quite so rosy when it comes to the strong interaction, but great progress has been made towards unifying the description of this fundamental force with the electroweak theory. By the 1980s, when the basic quark model of protons and neutrons was well established (once again, Feynman was involved in this; see Chapter 10), physicists had been so impressed by the success of QED and the electroweak theory that they deliberately set out to explain the strong force in similar terms. The picture that emerges is that protons and neutrons are each composed of three fundamental particles, called quarks, bound together by the exchange of particles which do the same work as photons do in QED and intermediate vector bosons do in weak interactions. This makes quarks and leptons the truly fundamental building blocks of everyday matter. The strong force, as we see it operating between protons and neutrons, is then explained as a residue of the real strong force operating between quarks, the truly fundamental fourth interaction, alongside gravity, electromagnetism and the weak nuclear force.

Quarks come in several varieties, revealed by high-energy events in particle accelerators like those at Fermilab and at CERN. But happily for us only two varieties are needed to make protons and neutrons. These have been whimsically given the names 'up' and 'down'. Among their other properties, each up quark carries an electrical charge of $+\frac{2}{3}$, while each down quark carries an electrical charge of $-\frac{1}{3}$. A neutron consists of two down quarks and one up quark bound together by the strong interaction, and a proton consists of two up quarks and one down quark bound together by the strong interaction. On this picture, neutron decay actually involves the transformation of a down quark into an up quark, with the aid of an intermediate vector boson (a W particle) linking the transforming quark to an electron–antineutrino pair.

The particles which are exchanged between quarks and bind them tightly together are also given a whimsical but this time descriptive name – gluons. Gluons carry the strong force, in the same way that photons carry electromagnetic forces. They can do this because the quarks themselves have another kind of charge, as well as, and distinct from, their electrical charge. In order to distinguish this charge from electrical charge, and for want of a better name, it is called 'colour'. Unlike electrical charge, colour charge comes in three varieties, not two. Instead of just plus and minus charge, we have 'red', 'blue' and 'green' charge. This doesn't mean that quarks are 'really' coloured; it is just a type of label. Remember that the names plus and minus for electrical charge are themselves arbitrary conventions, and only pass without comment because we are so used to them. The two kinds of electrical charge could themselves have been dubbed 'red' and 'blue', or (more plausibly) 'up' and 'down' when they were discovered. The property represented by the 'redness', 'greenness' or 'blueness' of a quark could itself just as easily have been dubbed eeny, meeny and miny, or anything else you like. But calling this property colour charge does have one neat side benefit; it means that the theory of how the strong force works, by exchanging colour charge between quarks with the aid of 'coloured' gluons, can be called quantum chromodynamics, or QCD for short.

QCD is an extremely successful theory in its own right. But it is a more complicated and mathematically hairy theory because more kinds of particles and varieties of 'charge' are involved. A major problem is that the higher-order terms in the calculation of things like the magnetic moment of, in this case, the proton are much more important in QCD than in QED. In QED, allowing for just three virtual photons being emitted and reabsorbed gets you close to the experimental number, but in QCD terms with six junctions involving gluons would have to be calculated to get anything like the same accuracy. The experiments are pretty accurate – they tell you that the magnetic moment of the proton is 2.79275. But the best calculations yet carried out with QCD 'only' give a value of 2.7, with an error of ± 0.3.

In *QED*, Feynman dismissed this as pretty poor – an error of 10 per cent, 10,000 times less accurate than experiment. In fact, the result is pretty impressive, as long as we don't use the yardstick of the superb accuracy of QED itself, and shows just how good QCD really is.

Nevertheless, partly because of these problems, it has proved very difficult to make QCD exactly fit the QED template, and it has not yet proved possible to unite QCD and electroweak theory into one single mathematical package, a so-called 'Grand Unified Theory', or GUT. Even if that can be achieved, there will still be the problem of bringing in gravity as well, to make a unified 'Theory of Everything', or TOE (more of this in Chapter 14). But in spite of its imperfections, QCD is a pretty good theory; it just isn't quite as good as QED itself. And all of the success of QCD in explaining the workings of the world at the level of quarks and gluons depends directly and explicitly on the applications of the QED template to this deeper level of the structure of matter – not just QED, but specifically Feynman's formulation of QED and the use of Feynman diagrams. The tools that Feynman developed half a century ago are still the tools being used by theoretical physicists at the cutting edge of research today.

This is not without its little irony, because Feynman himself was never convinced that he really had said the last word in quantum electrodynamics. In particular, like Dirac, he was never entirely happy with renormalization, which he described in his Nobel lecture as 'a way to sweep the difficulties of the divergences of electrodynamics under the rug'. In *QED*, he used more typical Feynman language to describe renormalization: 'It is what I would call a dippy process!'

Dippy or not, it worked. Feynman's version of QED was the last word on quantum theory in 1949, and it is still the last word today. Feynman's last two great papers on quantum theory were published in 1951, but everything had been worked out by the end of 1948. As Feynman said later,[5] he had 'disgorged myself of all the things I had thought about in the context of quantum electrodynamics . . . I had completed the project on quantum electrodynamics. I didn't have anything else remaining that required publishing. In these two papers, I put everything that I had done and thought should be published on the subject. And that was the end of my published work in this field.'

By the middle of 1951, Feynman was 33 years old. He could have rested on his laurels, led a quiet life as a professor at Cornell, never done any more research, and still he would have won the Nobel Prize and gone down in history as one of the greatest physicists of the twentieth century, 'another Dirac'. But that wasn't Feynman's way. By now, he was becoming restless, finding Cornell not as congenial a working environment as he had hoped, and finding new fields of physics to conquer. It was time to move on, both physically and as a physicist.

Notes

1. See Bibliography. This book is a masterpiece of clarity, in the authentic Feynman voice, transcribed and edited by Ralph Leighton from a series of lectures by Feynman. We follow it closely in our description of Feynman's masterwork.
2. Mehra.
3. Quoted by Gleick.
4. If you want to know more about how this works, see John Gribbin, *In Search of Schrödinger's Cat.*
5. Mehra.

— 7 —

The legend of Richard Feynman

By the end of the 1940s, there were many reasons for Feynman to feel restless. Professionally, although he was in the process of achieving his greatest triumph, he was also just passing his thirtieth birthday, and must have been aware that very few great physicists have made major contributions to their craft after passing that landmark. Dirac himself, Feynman's hero, was a good example of a physicist who achieved much in his twenties, and very little of any real importance thereafter. Indeed, there is a piece of doggerel, sometimes ascribed to Dirac, which makes the point forcefully:[1]

> Age is, of course, a fever chill
> that every physicist must fear.
> He's better dead than living still
> when once he's past his thirtieth year.

There have been very few exceptions to this rule. Erwin Schrödinger was 39 when he made his greatest contribution to science, the wave version of quantum mechanics. But that was very much a special case, since Schrödinger was deliberately harking back to old ideas about waves, trying to rescue quantum mechanics from the mess it seemed to have got into and return it to the comfortable physics he had learned in his youth. In that sense, it was very much the work of a (relatively) old man, looking backward rather than forward. A more relevant exception was Einstein, who continued to make significant and forward-looking contributions to quantum theory until well into his forties – but even the 30-year-old Dick Feynman might have stopped short of regarding himself as another Einstein.

There was also a problem with his social life. As a young, good-looking, charming and extrovert professor at Cornell, Feynman had achieved considerable success with women. By the standards of the late 1940s, he had achieved (if that is the right word) a reputation as a ladykiller which, with hindsight, can be seen as overcompensation for the loss of Arline. One of his most successful ploys was to hang out in the student union (Willard Straight Hall), drinking coffee and offering to help pretty girls who were having difficulty with their physics homework. In a typical Feynman anecdote, where the truth (or at least, part of the truth) is made palatable with layers of humour, he later told a colleague that he had decided to leave Cornell 'when he tried that routine on a coed and she said, "I know who you are. You're not a student, you're Dick Feynman." '[2] Fame, it seemed, did have its drawbacks.

More seriously, by hanging out with the students Feynman came to appreciate that a lot of what was being taught at Cornell was what he regarded as dopey stuff. This might not have been the sort of thing you would notice when working flat out on a theory like QED, but once the pressure eased and he had more time to take stock, it became a major nuisance. To someone who regarded English literature and philosophy as distinctly dippy subjects, it was utterly bizarre to find that a student could spend four years studying home economics or hotel management (he had first-hand experience of the hotel business, after all) and end up with a degree that was, on the face of it, as good as a degree in physics. There were exceptions – the physics school itself, of course, and some of the other scientific work being done at Cornell. But it was rare for Feynman to find anyone outside his own field with whom he could enjoy an intellectual conversation about their work. He met with what he described to Mehra as a general 'dopiness' among the students and faculty, a 'low-level baloney', quite different from his recollections of his own student days at MIT and Princeton. Not that he was against dopiness *per se* – just that 'It's not all right if you are talking to students and professors. That bothered me enormously.'[3]

And then there was the weather. Cornell is in upstate New York, in the small town of Ithaca, and it gets cold there in winter. In *Surely You're Joking*, Feynman graphically describes the hassle of driving in snow, stopping and fitting snow chains to the wheels with frozen fingers, 'and your hand's hurting, and the damn thing's not going down – well, I remember that *that* was the *moment* when I decided that *this* is *insane*; there must be a part of the world that doesn't have this problem'.

One option was a move to South America. Feynman had been intrigued by the possibility after picking up a hitchhiker who told him how interesting it was, and suggested that he might go there.[4] It wasn't just the weather that appealed. This was in the early years of the Cold War, when many of Feynman's former Los Alamos colleagues were, he knew, involved in work on the hydrogen bomb, and he still felt that nuclear war was inevitable (which was perhaps also a factor in his wilder adventures). It is hard to appreciate today just how seriously the threat was taken in those days, right through the 1950s and into the 1960s, and Feynman was by no means alone in thinking that South America might be a safer place to settle down than the United States. He even went so far as to learn Spanish, in preparation for a trip south, because it was the most widely spoken language in South America. But that turned out to be a mistake.

Early in 1949, Feynman met a Brazilian physicist, Jaime Tiomno, who was visiting Princeton. When Tiomno heard of Feynman's vague plans to visit South America, he offered to arrange for Feynman to spend part of the summer at the Brazilian Centre for Research in Physics, in Rio de Janeiro. The offer was irresistible, but it meant Feynman had to take a crash course converting his Spanish into Portuguese in time for the trip.

The six-week visit to Rio, in July and August 1949, was a huge success. Feynman's first encounter with the relaxed lifestyle came when he landed at Recife to change planes, and was met by representatives of the Centre. His onward flight was cancelled, and the next scheduled flight, 48 hours later, would not get him to Rio until the following Tuesday, a day after he was supposed to take up his summer post.

> I got all upset. 'Maybe there's a cargo plane. I'll travel in a cargo plane,' I said.
>
> 'Professor!' they said. 'It's really quite nice here in Recife. We'll show you around. Why don't you relax – you're in *Brazil*!'[5]

In Rio, Feynman taught physics in the mornings (lecturing in what he called ' "Feynman's Portuguese," which I knew couldn't be the same as real Portuguese, because I could understand what I was saying, while I couldn't understand what the people in the street were saying') and relaxed on the beach in the afternoons. There were other physicists to talk to, including Cecile Morette who was visiting the Centre from France, and lots of pretty girls (one of whom actually came back to Ithaca with him, but stayed for only a short time). Rio was definitely Dick Feynman's kind of place.

Returning from there to Cornell in the autumn of 1949, with the prospect of another New York winter ahead, may have helped to focus Feynman's mind on a more permanent move to warmer climes. By now Robert Bacher, another member of the old Los Alamos team, was head of the Division of Physical Sciences at the California Institute of Technology, and he invited Feynman to give a series of lectures at Caltech between January and March 1950. Feynman leapt at the opportunity to escape from the New York winter, and while he was in Caltech Bacher sounded him out about making the move on a permanent basis. Caltech had everything going for it – the climate in Pasadena was a distinct improvement, but most of all the place lacked dopiness. There were no home economics students there, but there were plenty of good scientists, everything from astronomers to zoologists. Caltech, too, was Dick Feynman's kind of place.

The only thing that made the prospect of the move difficult was that it would mean leaving Bethe, Feynman's mentor both at Los Alamos and in the difficult years at Cornell getting his theory of quantum electrodynamics established. Once again, Feynman was in demand, and when Cornell learned he was thinking of moving they made him a better offer, only for Caltech to increase their offer. Feynman really was undecided (around this time, he also asked the Centre in Rio if there was any chance of a permanent post there), until in the spring of 1950 Caltech found the ultimate sweetener. If Feynman stayed at Cornell, he would be entitled to a sabbatical year, which would give him a chance to go back to Brazil for an extended stay. Caltech said, OK, come here and you can still have the year off to go to Brazil, at our expense instead of Cornell's. That clinched it. Feynman agreed that he would take up the appointment at Caltech in the autumn of 1950, with the promise that he could spend the academic year of 1951–52 in Rio.

Before that, he made his first trip to Europe, in April 1950, to attend an international scientific gathering in Paris, and went on briefly to Zurich, where he lectured at Einstein's old school, the ETH (Federal Institute of Technology). Paris, also, turned out to be Dick Feynman's kind of place: 'I had met several of the girls who were dancing at the Lido in Paris, at Las Vegas. I watched rehearsals at the Lido, went backstage, and had all kinds of fun.'[6]

Las Vegas? How come Feynman knew the showgirls from Las Vegas? As he recounts in *Surely You're Joking*, most summers while he was at Cornell he used to set off west in his car, heading for the Pacific Ocean. 'But, for various reasons, I would always get stuck somewhere – usually in Las Vegas.' The 'various reasons' came down to having a good time, not just by participating in the usual activities in a place like Las Vegas, but by watching how the people there, and the whole set-up, operated. One way and another, by the time he was 30 Feynman was well set for the lifestyle that would continue for much of his next decade, teaching and researching at Caltech, travelling the world to attend scientific meetings, and having fun at the beach or in places like Las Vegas. With his reputation as a scientist already established, this was the period in which the legend of Dick Feynman the scientific playboy arose, and from which many of his own anecdotes and reminiscences are drawn. And the first big adventure of his new life out west was the sabbatical year spent in Brazil.

Feynman didn't really settle in Pasadena during his first year out west, the academic year 1950–51. He still wasn't sure that Caltech was going to be a permanent home, and still thought he might move back east, or (more likely) find a way to persuade the Brazilians to offer him a permanent post. So he stayed for the entire year at the faculty club on campus, the Athenaeum, and deliberately didn't try to put down any roots. But neither was Caltech a clear-cut break with the past, at least as far as his personal life was concerned. Among the women Feynman had dated at Cornell was Mary Louise Bell, a student of art history who came from Neodesha, Kansas. Mary Lou, as she was known, was something of a blonde bombshell, a few months older than Dick. She was the kind of woman his friends weren't at all surprised to see him involved with in a short-term relationship, but they would soon be dumbfounded when she became his second wife. Although Dick and Mary Lou often quarrelled, and had, as we shall see, ultimately incompatible personalities and ideas about how to live, she had one thing going for her (apart from her looks) – she wasn't dippy, and knew a great deal about Mexican art, which also fascinated Dick. And although they had met at Cornell, when Dick moved to Pasadena she turned out to be living near the University of California, Los Angeles (UCLA) in nearby Westwood.

Even so, the relationship didn't develop particularly seriously during Feynman's first year at Caltech, and he left for Brazil in the summer of 1951 very much a free agent. He was at the Centre for Research in Physics, this time for 10 months, from August 1951 to June 1952, funded partly by Caltech and partly by a programme of the US State Department. He stayed at the Miramar Palace Hotel in Copacabana, overlooking the beach; the hotel was also favoured by airline crews from Pan American during their stopovers in Rio, and Feynman soon became a regular member of their crowd, socializing with the stewardesses and getting through some serious drinking with them in the bars. But one day, in the middle of the afternoon, he realized that this was getting to be more than a social habit.

> I was walking along the sidewalk opposite the beach at Copacabana past a bar. I suddenly got this treMENdous, strong feeling: 'That's *just* what I want; that'll fit just right. I'd just love to have a drink right now!'
>
> I started to walk into the bar, and I suddenly thought to myself, 'Wait a minute! It's the middle of the afternoon. There's nobody here. There's no social reason to drink. Why do you have such a terribly strong feeling that you *have* to have a drink?' – and I got scared.
>
> I never drank again . . . You see, I get such fun out of *thinking* that I don't want to destroy this most pleasant machine that makes life such a big kick.[7]

In Brazil, using that wonderful thinking machine, Feynman taught courses in the mathematical methods of physics, and on electricity and magnetism. He carried out research into the nature of particles known as mesons, in collaboration with Leite Lopes, one of his Brazilian colleagues. He began to think seriously about the puzzling properties of liquid helium (more of this in Chapter 8). He also worked on the theory of the structure of the nuclei of some of the lighter elements.

For this last piece of work he needed to compare the theory with experimental data, just as the theory of quantum electrodynamics developed by making comparisons with experiments such as the measurement of the Lamb shift or the measurement of the magnetic moment of the electron. The way he kept up to date with the latest experiments, being carried out in the Kellogg Radiation Laboratory at Caltech, highlights the way the world has changed, at least as far as communications are concerned, since 1951. Today, a scientist anywhere in the world wishing to get the latest news from another scientist anywhere else in the world would use e-mail and the Internet. You'd get the latest data delivered right into your computer, ready to analyse, without even the chore of keying the numbers in for yourself. In 1951, though, even telephone communication between the United States and Brazil was unreliable and inconvenient. So Feynman communicated with Caltech with the aid of amateur radio operators. About once a week, he would go over to the house of a ham operator in Rio, who would contact a ham in Pasadena, who would pass on the latest news from the Kellogg Lab. 'The contact I had with Caltech by ham radio was', said Feynman, 'very effective and useful to me.'[8]

The contact he had with the students in Brazil was less effective, as he explains in *Surely You're Joking*, because the students had been taught how to learn by rote from books and lectures, without any understanding of what physics was really all about. He explains how the students could recite the definition of Brewster's Angle, which tells you (if you understand it) that when light is reflected off the sea it becomes polarized. But they were astonished, when he asked them to look at the sea through a polarizing filter, to discover that light reflected from the sea is polarized! There was no contact between their book learning and the real world. It was just like Melville's story about the 'Spencer's warbler'. The students had learned a list of facts, but had no idea what the facts really meant, and no understanding of how to discover new facts.

At the end of his visit, Feynman gave a talk explaining this problem at the core of Brazilian science teaching. Back at Caltech, he wrote this up for *Engineering and Science*, the Caltech magazine; the article stands today as an explanation of what physics, and physics teaching, is all about:

> Science is a way to teach how something gets known, what is not known, to what extent things *are* known (for nothing is known absolutely), how to handle doubt and uncertainty, what the rules of evidence are, how to think about things so that judgements can be made, how to distinguish truth from fraud, from show . . . in learning science you learn to handle by trial and error, to develop a spirit of invention and of free inquiry which is of tremendous value far beyond science. One learns to ask oneself: 'Is there a better way to do it?'[9]

You can see how this spirit of free inquiry, learning by trial and error, and all the rest, suffused Feynman's life. One of his favourite anecdotes concerns the way he learned, during his time in Rio, to play in a samba band, developing a skilful technique with a small percussion instrument called the *frigideira*, a round metal plate on a handle, about 6 inches across and looking like a little frying pan, that you beat with a little metal stick. He applied himself to this in the same way that he applied himself to physics, and for the same reason – because it was fun. That, deep down, was probably the reason for the gulf between Feynman and the students in Brazil. They were studying because it was the sober, sensible thing to do in order to get on in the system and get a job. He studied physics for the pleasure it gave him.

But there was still a gap in his life. Nineteen fifty-two marked the tenth anniversary of his marriage to Arline, and there was a gap that could not be filled by all the short-term relationships. One day, near the end of his stay in Rio, Feynman took one of the air hostesses to the museum. He was showing her the Egyptian section, explaining everything as they went along, 'and I thought to myself, "You know where you learned all that stuff? From Mary Lou" – and I got lonely for her.'[10]

He got so lonely for her, indeed, that he proposed to her by letter. 'Somebody who's wise could have told me that was dangerous: when you're away and you've got nothing but paper, and you're feeling lonely, you remember all the good things, and you can't remember the reasons you had the arguments.' Mary Lou, who was by now teaching at Michigan State University, accepted; but more arguments, about things like furniture and setting up home, continued by letter even before Dick got back to California.

Feynman returned from Brazil in June 1952, and made the commitment to stay at Caltech. The marriage took place with almost indecent haste, on 28 June 1952. Of course, the timing fitted in with the cycle of the academic year, and gave the couple the chance of a summer honeymoon visiting Mexico and Guatemala. It still looks odd, though, that it should have been exactly one day short of the tenth anniversary of his wedding to Arline, and suggests that Feynman was, consciously or subconsciously, trying to get his life sorted out into some sort of settled order before that landmark. The couple settled in Altadena, just to the north of Pasadena – but settled really isn't the right word, whatever Dick's subconscious may have been hoping. Mary Lou liked the idea of being a real professor's wife, and wanted Richard to act like a real professor, including wearing a jacket and tie and all the stifling social niceties that that implies. When they both visited Brazil in the summer of 1953, when Feynman spent a few weeks at the Centre working with his old friends, his Brazilian friends were amused to notice that he came in fully dressed up in necktie and jacket, until one day he turned up in his shirtsleeves. That was the day that Mary Lou had left Rio.[11] She had no time for scientists, and actively tried to cut him off from social contact with them by 'forgetting' about invitations. On one widely reported occasion, Feynman missed a chance to meet up with Niels Bohr on a rare visit to Pasadena, Mary Lou only mentioning to him, after it was too late to accept, that he had been invited to have dinner with 'some old bore'.[12] And when she could be persuaded to go along with Dick to a party, she made it quite clear that she disapproved. She would sit quietly in a corner at first, but would get increasingly annoyed when Dick went into his drunk routine – since he had given up alcohol, Feynman usually used to play drunk at parties, adjusting his behaviour smoothly to match the increasing alcohol intake of everyone else there. Soon, the reprimands would start coming from Mary Lou's corner: 'Richard. Richard! Stop that! You're acting like a fool, stop that!'[13]

Somehow the marriage lasted for four years, until the summer of 1956; but the writing had been on the wall from the beginning. Perhaps the best thing about it was the way it ended, with Dick agreeing to admit to extreme cruelty as grounds for the divorce. Since he wasn't actually a wife-beater, they had to dream up some way of making this stand up in court, and the novelty of the excuse they came up with caught the fancy of the press. The basis of this extreme cruelty was described in the *Los Angeles Times* on 18 July 1956, under the headline 'Beat Goes Sour: Calculus and African Drums Bring Divorce'. Mary Lou was quoted as testifying that her husband's bongo drumming made a terrible noise, and that he not only began working on calculus problems in his head as soon as he awoke, but 'did calculus while driving his car, while sitting in the living room and while lying in bed at night'. Extreme cruelty, indeed.

In the middle of this short-lived attempt to settle down, some time in the fall of 1954, Feynman once more, and for the last time, considered leaving Caltech. Although he doesn't say so in *Surely You're Joking*, the unsettled state of his marriage must have been a contributory factor, but the trigger was a really bad attack of smog. Conveniently forgetting how much he had hated the winter in upstate New York (as with Mary Lou, distance lent enchantment), he actually called Cornell and asked if he could have his old job back. They made encouraging noises. But the very next day, on his way in to work Feynman was met by a breathless colleague at Caltech, who came running up to Dick with the exciting news that Walter Baade, working at the Mount Wilson Observatory in the nearby San Gabriel Mountains, had found evidence that the Universe was much older than had previously been thought. Even before Feynman had got to his office, another colleague, Matt Meselson, came up and told him about a breakthrough he had just made in the study of DNA. Both were important, fundamental discoveries, at the cutting edge of science, from two widely different disciplines. Feynman realized that he would have to be crazy to leave such a place:

> And I realized, as I finally got to my office, that this is where I've got to be. Where people from all different fields of science would tell me stuff, and it was all exciting. It was exactly what I wanted, really.[14]

So he never did move back to Cornell, or on to anywhere else, in spite of offers. Nineteen fifty-four, halfway through the disastrous marriage to Mary Lou, the year in which Feynman received the prestigious Albert Einstein Award (not just prestigious – it brought with it $15,000 and a gold medal), was the year he finally made his own commitment to Caltech, and started to settle down, as far as Feynman could ever settle down.

It was easier to settle down at one permanent home base, of course, because he was in such demand to attend international conferences and to give guest lectures at other universities, not just in the United States but around the world. In September 1953, he visited Japan for the first time, for a meeting which took place partly in Tokyo and partly in Kyoto; Mary Lou stayed behind on this occasion. Typically, Feynman entered enthusiastically into the spirit of the adventure, learning some Japanese, practising eating with chopsticks before he left California, and insisting on staying in traditional Japanese-style hotels where he could absorb the atmosphere. He went back to Japan (this time with Mary Lou) in the summer of 1955, on a lecture tour of Japanese universities; in between, in March 1954, he visited the University of Chicago and gave a series of lectures as a guest professor. And he visited Europe on several occasions, as well as making trips back to Brazil – all officially working visits, quite separate from his real holidays.

There were, though, irritations associated with his growing fame. One of his most annoying encounters, to Feynman himself, was with the US National Academy of Sciences, which elected him a member in April 1954. He had never heard of the organization, which made no significant contribution to science, published what he discovered to be, when he looked at it, a distinctly second-rate journal, and seemed to be nothing more than an honorary society, which existed chiefly for the incestuous purpose of deciding who else was grand enough to be allowed to join its ranks. He was persuaded that by refusing to accept membership he would embarrass many of his friends, and that it was better to accept quietly. But when he went along to a meeting of the society, giving them a fair chance, it was deeply depressing. The main topic of conversation was who else should be elected to this honorary society, while the experiments that were reported were, in many cases, totally unscientific. Feynman was particularly unimpressed by an experiment in which rats had been observed drowning, with their efforts to survive being timed and monitored – a cruel and needless experiment with no scientific value.[15] He eventually resigned from the NAS, but without making a great deal of fuss.

After the divorce from Mary Lou in 1956, he established a nice routine. He had his research to do at Caltech, ample opportunity to visit research centres around the world, and he could return to his old haunts in Las Vegas for relaxation. By contrast with the hassles during his ill-fated second marriage, it was a good life, and he found his second wind as a bachelor, at least on the surface. By now, though, he was in his late thirties, and the gap in his life that Mary Lou had so conspicuously failed to fill was still there.

In the summer of 1958, Feynman was in Europe again, on a visit culminating in a contribution to the United Nations 'Atoms for Peace' conference in Geneva, in the first two weeks of September. He was on his own, and rather than stay in a big hotel with the other scientists and dignitaries attending the conference, he sought out a little place called the Hotel City – the same kind of little place that he had stayed in with Freeman Dyson, the night they were marooned by floods in Vinita. The 'hotel' was delighted to have a real guest – especially one who received telephone calls from the UN.

As told by Feynman, it was all another great prank.[16] Although he had just turned 40, he'd had a highly productive few years as a scientist (more of this in Chapter 8), and he seemed to be as happy as ever. But maybe his subconscious was at work again when, during a break from the conference, he struck up a conversation with a young woman in a blue polka dot bikini on the beach of Lake Geneva. She turned out to be Gweneth Howarth, a 24-year-old Englishwoman from a small village in Yorkshire, with a streak of adventure that rivalled Feynman's own.

Gweneth had had a routine upbringing in Yorkshire, becoming a school librarian and facing the prospect of a routine, humdrum life. Her sister Jacqueline recalls[17] that the children had a happy childhood in a close-knit family community, even though their mother had died when Gweneth was only six weeks old. Their father brought up the two girls with the aid of four aunts, and they enjoyed a comfortable childhood filled with music and dancing lessons, country walks, a succession of pet animals and all the benefits of country life. Gweneth was particularly fond of animals, and interested in gardening (much later, she became a landscape gardener).

Both girls passed the examination for the local grammar school, and after school Gweneth trained as a librarian, seen as a good occupation at that time (in the 1950s) for an independent, lively minded woman. Jacqueline remembers that the two sisters both had wanderlust, and that although they had not escaped to any Feynman-type adventures, they both travelled abroad on holiday much more often than was common at the time (Jacqueline still has this wanderlust; when we interviewed her she had just returned from Goa). But Gweneth was strongly attached to her surroundings and her family, and also formed a longstanding relationship with a boy from Halifax. It was only after this relationship ended that she decided to see the world, and set off intending to go to Australia.

In 1958 she quit work and bought a one-way ticket to Geneva, the first leg of a planned round-the-world trip. At one level, the family were surprised at the decision; but they also accepted that Gweneth was always her own woman, and could not be deterred from anything she had set her mind on. In an article for *Engineering and Science*, she told how her friends reacted to the news – some said, 'you're mad' while others said, 'I'd like to do it too', but nobody else did do it.[18] She took only a little money with her, so that she would not be tempted to buy a ticket straight back home. She had made no arrangements to work in Switzerland, but she would have to find work if only to get the fare home and, she reasoned, if she started working to earn fare money she would be able to earn enough to live on *without* coming home. Then, when she had organized her finances, she planned to carry on travelling around the world. At the time she met Feynman, she was working as an *au pair* for her keep plus pocket money, and had only three hours free on Thursday afternoon and three hours free on Sunday afternoon. It was in one of these rare periods to herself that she met Richard.

When Feynman learned of her circumstances, her plans to travel the world, and how little she was being paid (the equivalent of $25 per month), he suggested that she came to California. He needed someone to keep house for him – a maid, as he put it – and he could afford to pay her $20 a week, not $25 a month, plus her keep. At first, Gweneth didn't take the notion seriously. She had two boyfriends in Geneva, and as far as she had any plans they involved going to Australia for a couple of years; she had no particular fancy to visit the United States.[19] Feynman apologized for the brashness of his proposal. But she got on well with Richard; before he left Geneva she agreed to consider his offer, and they exchanged addresses.

By November, Gweneth had decided to take up Richard's offer, and began the process of sorting out an immigration visa for the United States. This involved a lot of tedious bureaucracy. In order to enter the United States to work, Gweneth would need a sponsor – somebody who would undertake to look after her financially if need be, until the work materialized, or if the job fell through. Feynman's lawyer told him that it would be a bad idea for him to be the sponsor himself, for a young woman who would be living in his own house, because he might fall foul of legislation concerning the transportation of women for immoral purposes. So Richard had to persuade a physicist friend, Matthew Sands, to act as sponsor, on the understanding between them that if Gweneth really did need financial help it would come from Feynman, not Sands. Eventually, the visa came through, and Gweneth, now 25, arrived in Altadena in June 1959.

Gweneth's family had missed her when she went to Geneva, and were very worried when she announced that she was off to California that they would never see her again. At the end of the 1950s, even crossing the Atlantic, let alone the North American continent as well, was a big adventure.

Everything was as Feynman had promised. He really did need someone to look after him. In an interview with Gleick, Gweneth later told how Dick had reduced his wardrobe to five identical pairs of shoes, identical dark blue suits, and white shirts that he wore with open collar. He had no TV or radio, and always kept keys, tickets and loose change in the same pockets, so that he would never have to think where they were.

He lived in the front part of the house, and she had her own room at the back. 'People in my hometown did not have the gumption to do something like going to Geneva or to Pasadena', she said in the *Engineering and Science* article. 'It worked fine.'

At first, Feynman kept quiet about his new housekeeper. Scarcely anyone (except, of course, Matthew Sands and his wife) knew she was there. Then, colleagues noticed that Dick was going home for lunch, and soon there was gossip in the Athenaeum about Feynman living with a woman.[20] In fact, at first, in spite of Dick's reputation as a womanizer, their relationship really was as it had been described in Gweneth's immigration papers. She had no intention of marrying him. 'I had boyfriends here; I had a marvelous time. I would date Richard from time to time. Until suddenly, out of the blue, he proposed. I was never more surprised in my life.'[21]

On Feynman's side, his proposal in the spring of 1960 wasn't a sudden decision at all. He later gave his version of the story to Leighton.[22] He had realized how happy he was long before he made his proposal, and set himself a deadline, several weeks ahead, to see if his feelings changed. He decided to propose if he still felt the same way when the day he had set had arrived. The evening before the day he had chosen, he was so excited that he couldn't wait, and kept Gweneth up on a pretext until midnight, so that he could ask her to marry him as early as possible without breaking his promise to himself to wait until that day. Gweneth was a match for him, though. She said she had to sleep on it before reaching a decision, and made him wait until the next morning before giving him her answer.

They married on 24 September 1960, when he was 42 and she was 26, and stayed married for life. Jacqueline and her family did not travel to California for the wedding, because their son Christopher was too young at the time. They first travelled there in 1966, the year after Richard won the Nobel Prize. But from the very first, Gweneth (often with Richard) came back every year to walk in the Yorkshire Dales and visit her family, and after the first trip to California Jacqueline and her family also frequently made the trip to the West Coast, with Richard becoming part of their family. Richard and Gweneth's son Carl was born in 1962, and in 1968 they adopted a baby daughter, Michelle. Feynman had found true happiness as a family man at last, and settled easily into the role of father figure not just to his own family but to a rising generation of physicists around the world. According to Willy Fowler,[23] although everyone at Caltech had been impressed by Feynman in the 1950s, recognizing him as 'the smartest and wisest guy in the physics division', he was far from easy to get on with. But 'Feynman changed after his marriage to Gweneth. He became a much nicer guy. She was just such a sweet person; it was just the opposite with Mary Lou, who was "very strange". Mary Lou antagonized everybody; everybody was relieved when Feynman divorced her. When Feynman married Gweneth we all wondered how it would be; it turned out to be wonderful.'

In amongst the turmoil of his personal life in the 1950s, though, Feynman had completed two major pieces of work in physics, as well as several lesser contributions.

Notes

1. See Helge Kragh, *Dirac* (Cambridge University Press, 1989). Kragh believes the attribution of the verse to Dirac to be apocryphal, but it aptly sums up the feelings of many physicists about their creativity.
2. Michael Cohen, in *Most of the Good Stuff.*
3. Mehra.
4. *Surely You're Joking.*
5. *Surely You're Joking.*
6. Mehra.
7. *Surely You're Joking.*
8. *Surely You're Joking.*
9. *Engineering and Science*, Caltech, November 1953.
10. *Surely You're Joking.*
11. Leite Lopes, reported by Mehra.
12. Mehra.
13. Albert Hibbs, as told to Mehra.
14. *Surely You're Joking.* One reason why it was possible to meet such exciting people at Caltech was because Caltech itself was so small. Even by the early 1960s, there were roughly equal numbers of undergraduates, graduate students and faculty there – about 600 of each.
15. Mehra.
16. *What Do You Care.*
17. Interview with MG, February 1996; Jacqueline Howarth is now Jacqueline Shaw.

18. Gweneth Feynman, 'The Life of a Nobel Wife', *Engineering and Science*, March–April 1977.
19. See note 18.
20. Albert Hibbs, as told to Mehra.
21. See note 18.
22. Comment to JG, December 1995.
23. Mehra.

— 8 —

Supercool science

Feynman had become interested in the peculiar behaviour of liquid helium while he was still at Cornell, but he had been too busy completing his version of QED to devote any real effort to the puzzle. It was, though, a natural for him to take up once he got settled at Caltech in the early 1950s. It was a fundamental problem in physics, involving the quantum properties of particles, that he was able to tackle using his special insight into the behaviour of nature, seeing right to the heart of the problem and avoiding the thickets of mathematical complexity with which everybody else had surrounded the problem.

In order to liquefy helium at all, you have to achieve really low temperatures. The lowest temperature it is possible to reach, even in principle, is −273.16° Celsius, defined as zero on the Kelvin (K) scale. This 'absolute zero' is the temperature at which each particle has the minimum amount of energy which it is allowed to possess by the quantum rules. In a sense, this is an example of quantum uncertainty at work. If a particle had zero energy, it would be completely at rest, in one place, and not going anywhere. So there would be no uncertainty about its position and its momentum. In order for there to be uncertainty, it must always have at least a little energy so that it can jiggle about in different ways. Helium only condenses from a gas to form a liquid at a temperature of 5.2K, where the amount of jiggling it can do is already getting close to the quantum minimum. But what it does with the little energy it has can be spectacular.

Helium was first liquefied by the Dutch physicist Kamerlingh Onnes in 1908, and in further experiments he pressed on to temperatures even lower than 5K. In 1911, he discovered that something very peculiar happens to liquid helium at a temperature of just 2.2K; at about the same time, he discovered the phenomenon of superconductivity, the complete disappearance of electrical resistance in some metals when they are cooled to very low temperatures.

The first peculiar thing that happens to liquid helium when it is cooled below 2.2K is that it expands as it is cooled further, instead of contracting. Because of this, and other changes that occur at the same temperature, the liquid below this transition temperature became regarded as a separate 'phase' of helium, as distinct from liquid helium at higher temperatures as the liquid itself is from the gas. The liquid above 2.2K became known as helium I, and the liquid below 2.2K was dubbed helium II. The most impressive property of liquid helium II, established some time after Onnes' pioneering work, is that it is a superfluid – it can creep through tiny capillary tubes without seeming to meet any frictional resistance, and will even climb up the walls of a container to escape, or leak away through pores that are so small that gas cannot get through them.

The idea that was becoming accepted by the beginning of the 1950s, and which was producing the profusion of mathematical thickets surrounding the puzzle of superfluidity, was that below the critical temperature of 2.2K liquid helium II could be treated as if it were a mixture of two separate fluids. Part of the fluid seemed to have settled into the state it would be in at the absolute zero of temperature, 0K itself, with the minimum amount of energy in each helium atom. The rest was a 'normal' fluid. At 0K, the fluid would all be in the minimum quantum energy state, and at 2.2K it would be all 'normal', with the proportions varying smoothly in between.

The key to this interpretation of superfluidity is the way in which quantum entities behave, and it relied on treating whole helium atoms as if they were single quantum entities like electrons or photons – indeed, it specifically relied on treating them *exactly* as if they were photons.

Quantum entities come in two varieties, called fermions and bosons (after the physicists Enrico Fermi and Satyendra Bose). Fermions are what we are used to thinking of as particles, such as electrons; they each have an amount of quantum spin which is a half-integer – $\frac{1}{2}$, or $\frac{3}{2}$, or $\frac{5}{2}$, and so on. Bosons are what we are used to thinking of as waves, like photons; they each have zero or integer spin – 0, or 1, or 2, and so on. The important practical distinction between fermions and bosons is that no two fermions can exist in the same quantum state, while bosons can happily exist in the same quantum state as other bosons. This has implications, for example, for the structure of the atom. The electrons surrounding the nucleus of an atom must each be in a unique quantum state, sitting on different rungs of an energy level 'ladder'. Two electrons are allowed to sit on the bottom rung, because they can have opposite spins (one up, one down), but additional electrons have to sit, in some sense, successively further out from the nucleus in order to avoid being in the same state as one of these two inner electrons. The situation is slightly more complicated than we have made it sound, but the important point is that each electron has its own place, like the members of a theatre audience each with their own numbered seat in the auditorium. If it were not for the exclusivity of the fermions, all the electrons in an atom – *any* atom – would jostle together in the lowest energy state next to the nucleus, so all atoms would have more or less the same chemical properties and there would be none of the chemical complexity that makes the world so interesting and makes life possible.

Bosons obey different rules, and can pack together in the lowest energy state with other bosons. Rather than being placid theatre-goers sitting in their numbered seats, they are more like the enthusiastic fans at a rock concert, all crammed into the space in front of the stage together. There are other differences, which affect the way in which a box full of bosons – a boson gas – behaves, making its properties different from those of a fermion gas. One of the most dramatic discoveries of theoretical physics in the 1920s was that the behaviour of light can be entirely explained in terms of photons as particles obeying the rules appropriate for a boson gas, without invoking the idea of waves at all. Albert Einstein was involved in this work, and such a boson gas is sometimes referred to as a Bose–Einstein condensate. The two-component model of superfluid helium says that below 2.2K part of the fluid is behaving as a Bose–Einstein condensate (a boson gas), in the same way that photons behave, while the rest is behaving in the way that particles such as electrons behave (a fermion gas).

Feynman explained the superfluid behaviour of liquid helium in a series of 10 scientific papers (more than he published on QED) in a five-year period (1953–58) during the 1950s – many of them based on work carried out during the turmoil of his second marriage and its aftermath. As always, he started from first principles, largely ignoring the efforts other people had made to come to grips with the problem, and thinking about the behaviour of individual atoms in the fluid – the way they jiggled about, or slid past one another, or bounced off one another. He used the path integral approach, which turned out to be just as effective here as in QED or in classical optics, producing a theory that the physicist David Pines has described as 'that blend of magic, mathematical ingenuity and sophistication, and physical insight that is almost uniquely Feynman's'.[1] Pines also draws attention to the fact that the second paper in the series contains only a single equation, but leads the reader to certain conclusions about the behaviour of liquid helium, starting out from the fact that it is a Bose–Einstein condensate, through 'a series of closely reasoned arguments' alone. As well as establishing a satisfactory model of superfluidity, Feynman introduced a generation of condensed matter physicists to the use of Feynman diagrams and path integrals, making these techniques indispensable tools in that branch of physics.

Feynman also worked on the problem of superconductivity, but for once his insight let him down, and he was unable to come up with a satisfactory explanation of the phenomenon. And yet, even this failure has gone down in scientific folklore, because Feynman's response to it demonstrates another facet of his character, his scrupulous honesty in matters scientific. The problem was actually solved in 1957, by John Bardeen, Leon Cooper and Robert Schrieffer. Feynman was one of the first physicists to appreciate that their model (known as the BCS theory) really had solved the problem, and promptly abandoned his own efforts at explaining superconductivity while singing the praises of the BCS theory at every opportune occasion. But it was events at a conference held a year earlier, in 1956, that had impressed Schrieffer with Feynman's unique way of tackling physics. Schrieffer, as it happens, was the rapporteur for that meeting, and so paid close attention to all the talks. In an interview with Gleick, he has recalled how Feynman delivered a talk on two problems – the one he had solved (superfluidity), and the one that still baffled him (superconductivity). Schrieffer had never before heard a scientist describe publicly, in such loving detail, all the steps in a failed theory. Feynman's natural honesty helped others to avoid falling into the same traps he had fallen into, by signposting the danger areas, and showed clearly his ability to avoid deluding himself into thinking he was on the right trail when in fact he was barking up the wrong tree.

In 1972, the BCS team shared the Nobel Prize for Physics for their theory of superconductivity. Bardeen thereby made history, becoming the first person to win two Nobel Prizes in the same field, having already shared the physics prize with William Shockley and Walter Brattain in 1956, for their discovery of the transistor effect. With hindsight, it is hard to see Feynman's investigation of superfluidity as any less significant than the BCS theory of superconductivity, but as it happens when Lev Landau received his Nobel Prize in 1962 the citation specifically mentioned his work on the theory of liquid helium. In 1962, it was clear that Feynman's masterwork was QED, so there was probably no real consideration given to splitting that year's award between him and Landau; by 1972 (when the BCS team received their prize), it would have been too late, the prize for superfluidity having already been given. Otherwise, Feynman might well have shared Bardeen's double distinction.

Any disappointment Feynman may have felt in 1957 at his failure with the problem of superconductivity was, though, far outweighed by the joy he experienced that summer by making another Nobel-quality contribution to physics, in a completely different field again, the theory of the weak interaction.

Feynman had always been impressed by the beauty and power of Dirac's mathematical description of the electron, and had hankered after making a similar discovery. Such fundamental discoveries are very rare in physics; one of the few examples comparable to Dirac's equation for the electron would be Maxwell's equations of electromagnetism. So Feynman knew that this was a dream that would probably never be fulfilled. But he came close in 1957 – sufficiently close to satisfy himself that he had made a significant contribution – with his version of the theory of beta decay, the weak interaction process in which a nucleus (or an individual neutron) spits out an electron.

Feynman's close involvement with the theory of the weak interaction began at a conference held in Rochester, New York, in April 1956, and lasted just about 18 months. He had other things on his mind during that period – his divorce took place in the summer of 1956, and he was in the midst of his series of epic papers on superfluidity. But his attention was caught by a curious problem involving two types of particles, then known as theta and tau, which had first been discovered in cosmic rays. The puzzle was that in almost, but not quite, every way the theta and tau were identical – they had the same mass, and though the particles were unstable they each had the same lifetime, and so on. There was just one difference. When the theta particle decayed (through the weak interaction) it disintegrated into two particles in the family known as pions, while when the tau particle decayed it produced three pions. While the set of three pions had an amount of a property called parity equal to -1, the set of two pions had an amount of parity equal to $+1$. Assuming that no parity had been lost in the decay process, that meant that the theta and tau particles themselves had different parity, and so must really be different particles.

This notion of 'parity conservation' was a cherished belief of physicists, because it is related to the way things are reflected in a mirror. If parity is conserved, it means that nature, at a fundamental level, does not distinguish between left and right. If parity were *not* conserved, though, that would mean that the laws of physics would be different (perhaps only slightly different, but different) in Alice's looking-glass world. Several people were struggling with this problem, trying to find a way to allow the theta and tau to be the same particle while preserving parity conservation, in the period immediately before and after the 1956 Rochester meeting. They included a team of two Chinese-born American physicists – Chen Ning Yang (known to his friends as Frank) at the Institute for Advanced Study in Princeton and Tsung Dao Lee (known as T. D.) at Columbia University – and, working on his own, Murray Gell Mann, born in New York City in 1929, who had recently (in 1955) become a professor at Caltech, and would spend many years in the office next but one to Feynman, separated by the office of their secretary.

Feynman's roommate at the 1956 Rochester meeting was Martin

Block, an experimenter who felt diffident about challenging the cherished ideas of the theorists in public, but who, as Feynman recalled in *Surely You're Joking*, asked him one evening if it would really be so bad if parity were violated – if the theta and tau really were the same particle. Feynman thought this was a good question, and urged Block to ask the experts. But Block demurred, insisting that nobody would listen to him, and asking Feynman to pose the question:

> So the next day, at the meeting, when we were discussing the tau–theta puzzle, Oppenheimer said, 'We need to hear some new, wilder ideas about this problem.'
>
> So I got up and said, 'I'm asking this question for Martin Block: What would be the consequences if the parity rule was wrong?'[2]

Lee answered the question, but using complicated jargon which neither Feynman nor Block really understood. At least one other experimenter discussed with Feynman the possibility of carrying out an experiment to search for parity violation in other particle interactions, but didn't actually do the experiment. But Block later told Feynman that he had travelled home from the conference in the same plane as Lee, and had used the opportunity to press the case, arguing that at least the possibility was worth investigating.[3] Feynman went back to California to sort out his divorce and carry on his work on liquid helium. But Lee and Yang, who had already been working on the parity problem and may have been stimulated further by the discussions at the 1956 Rochester meeting, published a paper later that year looking at the whole situation of parity violation in weak interactions, discussing the theoretical implications, and suggesting experiments that could be carried out to test the idea. By the end of the year, Chien Shiung Wu, another researcher at Columbia University, had carried out one of the experiments proposed by Lee and Yang and had shown conclusively that parity *is* sometimes violated in weak interactions. Less than a year after that, in the autumn of 1957, Lee and Yang received the Nobel Prize for their work, one of the quickest such awards ever made. (Although Alfred Nobel actually specified that his prizes should be given for work carried out in the previous year, the rule is almost always broken.)

But although everybody knew, by the end of 1956, that parity was not conserved, and that therefore the theta and tau particles were the same thing (now called the kaon) decaying in two different ways, nobody had a satisfactory theory to describe such peculiar behaviour. The following April, at another of the annual Rochester meetings, Feynman took advantage of the opportunity to stay with his sister, Joan, who had completed her PhD in solid state physics and was living in nearby Syracuse. On this occasion, she was able to repay him handsomely for some of the sound advice he had given her, many years before, that had set her on the road to that PhD.

Richard had a copy of the paper Lee was to present to the 1957 Rochester meeting, and complained to Joan that he couldn't understand it.

> 'No,' she said, 'what you mean is *not* that you can't understand it, but that you didn't *invent* it. You didn't figure it out your *own* way, from hearing the clue. What you should do is imagine you're a student again, and take the paper upstairs, read every line of it, and check the equations. Then you'll understand it very easily.'[4]

Sounds familiar? Remember when Joan was 14, and Richard told her how to cope with the astronomy book he gave her – 'you start at the beginning and you read as far as you can, until you are lost. Then you start at the beginning again, and you keep working through until you can understand the whole book.'[5]

Richard took his sister's advice, and found that what he had thought to be difficult and incomprehensible was indeed 'very obvious and simple', once he got to grips with it. So much so, that he realized some old work he had done in another context could be applied to these problems, and made new predictions about the outcome of experiments involving the weak interaction. In a typical Feynman blitz, he worked through everything the same night, solving in his own way problems that others had been puzzling over for months. The theory of weak interactions that he came up with (which, of course, was based on the path integral approach) didn't quite work; while it made some clear-cut predictions, in other cases, including the archetypal example of neutron decay itself, it was still a bit messy. Nevertheless, it was progress. The next day, Feynman was able to persuade one of the scheduled speakers at the meeting, Ken Case, to give up five minutes of his time to allow Feynman to give a quick outline of his ideas to the conference. 'Then', as Feynman put it, 'I went to Brazil for the summer.'[6]

Nobody else worked like this. He had made a vital breakthrough, worked out the implications in a few hours and managed to summarize his discovery in five minutes. Then, instead of writing it up for publication, he went to Brazil. But Feynman was never worried about priority, or being beaten by other scientists, whether they were Euclid or Lee and Yang. Very often, he never bothered to publish his own work. Many times, a colleague would visit his office at Caltech to ask Dick's advice about a problem, only to find that he had solved it himself, long ago, and never even mentioned it to anyone. More than that, as Murray Slotnick had been disconcerted to discover at the 1949 meeting of the American Physical Society, usually Feynman had solved a much more general version of the problem.

This combination of skill with the physics and complete indifference to publication extended far outside the fields in which Feynman made his name. The astrophysicist Willy Fowler, who also worked at Caltech, had a favourite Feynman anecdote from the time, in the early 1960s, when quasars had just been discovered.[7] Fred Hoyle gave a seminar at Caltech suggesting that quasars might be supermassive stars, and was nonplussed when Feynman (an expert in quantum theory and superfluid flow, but not, as far as anyone was aware, in gravitational theory) stood up to say no, that was impossible, such a star would be gravitationally unstable. It turned out that Feynman had worked out a thorough treatment of the stability of supermassive stars, including a full account of the effects described by the General Theory of Relativity, years before, and essentially simply for his own amusement. According to Fowler, it ran to over a hundred pages of work, work which any astrophysicist would have been proud to have done, but he had simply never bothered to publish, having satisfied himself (the only audience he really wanted to impress) that it was right.

In fact, Fowler's anecdote gives a slightly distorted picture of the truth, because it was no secret that Feynman was interested in gravity – the surprise was how far that interest had taken him by the early 1960s, when quasars were discovered. He had actually attended one of the first conferences on the role of gravitation in physics, held at the University of North Carolina, Chapel Hill, in January 1957 (this was the occasion, described in *Surely You're Joking*, when he arrived late for the meeting, discovered that there were two campuses in North Carolina, and found his way to the right one by asking the cab driver if he had noticed the destination of a group of people 'talking to each other, not paying attention to where they were going, saying things to each other like "G-mu-nu. G-mu-nu."' The cab driver recognized the description of the physicists immediately, and took Feynman to the right campus[8]). So Feynman was actively involved in gravitational research even before the 1957 Rochester meeting where, at Joan's behest, he got to grips with Lee's paper.

After the Chapel Hill meeting, Feynman worked on gravitation for four or five years, trying to find a way to develop a quantum theory of gravity. He was especially interested in gravitational radiation, and was one of the first people to argue strongly that 'gravitons', the gravitational counterparts to photons, must exist. The search for gravitational radiation has as yet proved fruitless, but a new generation of detectors should be able to detect bursts of radiation from collapsing stars in the early part of the twenty-first century; appropriately, Caltech is one of the leading centres of this search today. But Feynman's own investigations of quantum gravity ran into a brick wall in the early 1960s. In July 1962, he attended a conference in Warsaw where he described the work he had done, and this work appeared in the proceedings of the conference, published in 1964.[9] Although progress has been slow in this field, Feynman's work (especially his use of the Lagrangian formalism) is still relevant today, as we shall see in Chapter 14. What really is remarkable, though (and this is surely the point of Fowler's story) is that Feynman carried out this work alongside his other investigations, including developing his theory of the weak interaction.

The big problem with Feynman's embryonic theory of the weak interaction, as he acknowledged at the 1957 Rochester meeting, was that it didn't work for neutron decay. It didn't work in a quite specific way, involving the types of virtual particles involved in the interactions. These particles, although ephemeral, are an essential ingredient in all modern theories of particle interactions (including quantum gravity!), and their own properties affect the way in which those interactions, including the weak interaction, take place. Some of the properties they possess, related to their spin and parity, were dubbed A, V, S and T (shorthand for 'axial', 'vector', 'scalar' and 'tensor', but the names don't really matter). Feynman's new description of the beta decay of neutrons said that it must involve V and A interactions, but the published experimental results on beta decay said that the process involved S and T interactions.

If he had stayed around and looked into this discrepancy with the experimenters, he might well have resolved it in the spring of 1957. But while Feynman was in Brazil, other physicists continued to puzzle over the problem. At the University of Rochester itself, home of the annual high-energy physics meetings, Robert Marshak (who had founded the Rochester gatherings in 1950) and his student George Sudarshan were coming round to the view that maybe beta decay could involve V and A interactions, after all; Murray Gell-Mann was thinking along similar lines at Caltech, and the three of them discussed the implications when Marshak and Sudarshan visited Caltech in July 1957, with Feynman still away in Brazil. Sudarshan, indeed, had already spent a lot of time on the problem before the April 1957 Rochester meeting – but as a student, he wasn't allowed to give a presentation there, and his supervisor, Marshak, was preoccupied with giving a major paper on another topic. Somehow, neither of them mentioned their work on the weak interaction in the discussion periods at that meeting.

On the way back from Rio, Feynman travelled via New York, and stopped off at Columbia University hoping to discuss the latest experimental results on the problem of the weak interaction with Wu. She wasn't there, but a colleague brought Feynman up to date with the situation – basically, it was still a mess. By the time he got back to Caltech, Gell-Mann was away on vacation, but Feynman went to talk the problem over with the experimenters. They agreed that the situation was hopelessly confused. 'It's so messed up,' they told him, 'Murray says it might even be V and A.'[10]

Feynman was electrified. If beta decay involved V and A interactions, not S and T, his theory was right after all! He calculated everything again, and it worked. At first, it seemed that a certain number calculated in accordance with his theory disagreed with experiment by 9 per cent; then he discovered that the number printed in the textbooks was wrong, and had since been revised by 7 per cent, in the right direction. The discrepancy was really only 2 per cent, pretty good for anything involving particle physics. In another all-night session, he calculated away, buoyed up by the euphoric feeling that he had made one of the truly fundamental discoveries in physics:

> I felt that it was the first time, and the only time, in my scientific career that I knew a law of nature that no one else knew. Now, it wasn't as beautiful a law as Dirac's or Maxwell's, but my equation for beta decay was a bit like that. It was the first time that I discovered a new law, rather than a more efficient method of calculating from someone else's theory.[11]

This is a rather self-deprecating way of referring to QED, given that in order to make his contribution there Feynman had found a completely new way to formulate quantum theory (and classical theory!) from first principles, but for whatever reasons the equation for beta decay was the discovery that he himself was most impressed by. 'Now,' he thought, 'I have completed myself.' And, just for once, he was sufficiently fired up to write the discovery up for publication immediately.

But things weren't quite that simple, and for once – the only time that it really mattered to him – Feynman's laid-back approach to the question of establishing priority was to cost him dear. Soon, Gell-Mann returned from vacation intending to write up his own version of the V and A theory of the weak interaction, and was somewhat miffed to find that Feynman had picked up what Gell-Mann regarded as his own ball and run off with it.

Pouring oil on potentially troubled waters, and anxious to avoid two rival papers on the same discovery by different authors coming out of Caltech at the same time, the head of the physics department at Caltech, Robert Bacher, urged Feynman and Gell-Mann to produce a joint paper, which they did. It was received by the *Physical Review* on 16 September 1957, and published in 1958 in less than six pages of the journal. It was a clear step forward in physics, in a sense giving the 'equation of the neutrino' in the way that Dirac had provided the 'equation of the electron'. It soon became (and remains) a widely quoted classic – much to the chagrin of Sudarshan and Marshak, who had written up their own version of the idea back in July, presented it at a meeting in Italy in the autumn of 1957, but only got into print in a journal (also the *Physical Review*) after Feynman and Gell-Mann. The result was that their work was unjustly seen as a 'me too' exercise. This came as a severe blow to Sudarshan, who was a young researcher who had just completed his first major piece of work, and realized he was unlikely to do anything as significant in the rest of his career; he never overcame the bitterness he felt about the way credit was apportioned to Feynman and Gell-Mann. In all fairness, though, the rule in science is that credit generally goes to the person who publishes first, and Sudarshan and Marshak had ample opportunity to get something into print between the 1957 Rochester conference and the end of that year; even when they did publish, their work was not so complete or elegant as the Feynman and Gell-Mann version. Feynman himself, though, always tried to give Sudarshan due credit, being careful always to refer to the work of Sudarshan and Marshak, as well as the Feynman and Gell-Mann paper, when discussing the theory of the weak interaction.[12]

But there was another lesson that Feynman learned from the experience of his work on the weak interaction. Why, if the interaction really involved V and A instead of S and T, had everybody been so certain that it was S and T? It turned out that all the experts had been quoting, some second or third hand, from one experiment. On advice from Robert Bacher, Feynman went to the library and looked up the paper which everybody quoted when saying that the weak interaction was S and T. He discovered that the conclusion was based on the positions of the last two data points on the edge of a graph, based on experimental measurements, plotted in that paper, 'and there's a principle that a point on the edge of the range of the data – the last point – isn't very good, because if it was, they'd have another point further along . . .'[13] Until then, he had 'never looked at the original data . . . Had I been a *good* physicist, when I thought of the original idea back at the Rochester Conference, I would have immediately looked [it] up . . . Since then I never pay any attention to anything by "experts." I calculate everything myself.'

This, maybe, explains why Feynman never got too upset that credit for his greatest discovery had to be shared. It was his own fault, and nobody else's, that he hadn't looked up the experimental data before going to Brazil, spotted the error and published his theory immediately. He could live with that; and, besides, he still knew that he had worked it all out by himself, even if some other people were almost as quick off the mark as he had been.

Maybe the theory of the weak interaction might, under other circumstances, have won somebody a Nobel Prize; the work is certainly of at least as high a standard as many of the achievements which have been honoured in that way. But there's a snag – one of the rules that is never broken is that the prize for a particular piece of work cannot be shared by more than three people. It is a ridiculous and arbitrary rule, but it means that the possibility of giving the award jointly to Feynman, Gell-Mann, Marshak and Sudarshan was never even discussed.

The theory of superfluidity and the theory of the weak interaction were Feynman's two great contributions to physics in the 1950s, and either of them would have been enough to establish the credentials of any ordinary physicist in the top rank of the profession, and provide him or her with sufficient kudos to last a lifetime. They pale in the Feynman legend only alongside the glorious brilliance of QED itself. And yet, alongside these two epic pieces of work, his domestic troubles, his visits to Las Vegas and other interesting places, his sometimes exotic social life and then his meeting and marriage with Gweneth,* in the 1950s Feynman also found time to make a few other contributions, by way of relaxation, in different fields of science and engineering.

In the mid-1950s, as if he didn't already have enough on his plate, Feynman got involved with the development of masers (forerunners to lasers) through the presence at Caltech of Robert Hellwarth, a maser specialist. In collaboration with a research student, Frank Vernon, they developed a simple way of calculating

*The social life and physics entwine delightfully here, since the reason Feynman was in Switzerland when he met Gweneth was that in 1958 the 'Rochester' conference had become peripatetic, uprooting itself and settling in Switzerland for a season.

problems involving masers and lasers, using a new kind of diagram as an easy way for engineers dealing with practical problems to come to grips with quantum mechanics; the work became one of Feynman's most cited contributions to physics,[14] and the chances are that the person who designed the laser in your CD player used the FVH technique in that work. Hellwarth moved on to work for the Hughes Aircraft Company, and through this connection Feynman began to give a series of lectures at Hughes, talking on any subject he liked. The lectures took place every week, on Wednesday, when Feynman was in California, and he enjoyed them so much that the tradition continued for the best part of 30 years.

Feynman was also interested in the new developments in molecular biology, the study of DNA, which carries the genetic message of life. One of the reasons why he decided to stay at Caltech, as we have seen, was the presence of biological researchers such as Max Delbrück on the campus, and the opportunity to keep bang up to date with developments in this field. In the second half of the 1950s, Feynman arranged with Delbrück and a younger biologist, Robert Edgar, that he would hang out in their department from time to time, acting like a graduate student in biology, being taught how to handle the biological material and given a small project to work on. This proved so interesting that when he became eligible for another sabbatical year, in 1959–60, Feynman spent it working on DNA studies at Caltech with Matt Meselson. He learned a great deal, without making any major contribution, and had an opportunity to get acquainted with many of the top researchers in the field. But the most pleasing aspect of his year in biology came through his duties as a teaching assistant. He taught first-year biology students, who had no idea who Dick Feynman was, the basic practical techniques of their trade, plus mathematics and statistics. At the end of the year, the students ranked him as the best teaching assistant they had encountered. 'I got a tremendous boost by obtaining the best score of all teaching assistants; even in biology, not my field, I could explain things clearly, and I was rather proud of it.'[15]

Feynman was, indeed, about to come into his prime as a great explainer. But his most memorable contribution to science during that sabbatical year in biology came from a one-off talk that he gave, at the end of December 1959, to the annual meeting of the

American Physical Society, which happened to be held in Caltech that year. The talk was titled 'There's Plenty of Room at the Bottom',[16] and it is hailed today as the first clear statement of the possibilities of nanotechnology – engineering on the scale of atoms and molecules.[17]

In the talk, Feynman threw out two challenges, offering a $1000 prize to the first person to solve each of them. One was to build a working electric motor that would fit inside a cube $\frac{1}{64}$ inches on each side. To his surprise (and consternation – he had made no arrangements to fund the prize, and paid up out of his own pocket) this was achieved by a local engineer, William McLellan, by November 1960. McLellan took his equipment along to show Feynman; it was in a large wooden box, and he has told how Feynman's eyes seemed to glaze over at the sight. Then, McLellan opened the box and took out a microscope with which to view his tiny motor. 'Uh-oh', Feynman said.[18]

The other prize was for anyone who could find a way to write small enough to get the entire *Encyclopaedia Britannica* on the head of a pin, a reduction of 25,000 times from its standard print size. On that scale, 'all of the information which mankind has ever recorded in books can be carried around in a pamphlet in your hand',[19] a pamphlet equivalent to 35 pages of the printed *Encyclopaedia Britannica*. This prize was claimed in 1985, by Tom Newman, a graduate student at Stanford University. He wrote out the first page of Charles Dickens' *A Tale of Two Cities* at the required scale, on the head of a pin, using a beam of electrons. The main problem he had before he could claim the prize was finding the text (using an electron microscope) after he had written it – the head of the pin was a huge empty space compared with the page of text inscribed on it. Ten years later, in 1995, scientists at the Los Alamos National Laboratory were literally copying the texts of entire books on the sides (not the heads) of steel pins measuring 25 by 2 millimetres, each capable of storing 2 Gigabytes of data in a permanent, readable form. What seemed like a wild flight of fancy at the end of 1959 became practical reality some 35 years later, with applications for data storage and retrieval anywhere that large amounts of information have to be stored in read-only form.

In the next century, such databases as the Library of Congress and the British Library really may be based on maintaining a collection of a few steel pins from which copies of any book ever written can be printed up on demand. All this was clearly foreseen by Feynman, and presented to the astonished gathering of physicists in 1959, like a magician pulling a rabbit out of a hat, by someone taking time off from being the best teaching assistant in the biology department at Caltech. Having finished his work on superfluidity and on the weak interaction, and having completed his biological sabbatical, at the beginning of the 1960s Feynman was without a major research problem to pursue over the next few years (except for his unpublished private investigations of gravitational theory), but had settled at last into a happy marriage. He was ideally placed to make the leap from being the best teacher at Caltech to being the best physics teacher in the world, reaching a wider audience than ever before (and pulling a few more rabbits out of the hat as he did so) with the books that, as Schwinger might have put it, brought Feynman's way of thinking about physics to the masses.

Notes

1. See *Most of the Good Stuff.*
2. *Surely You're Joking.*
3. Told by Feynman to Mehra.
4. *Surely You're Joking.*
5. See Joan Feynman's contribution to *No Ordinary Genius.*
6. *Surely You're Joking.*
7. Willy Fowler, conversation with JG, early 1970s.
8. Ralph Leighton, interview with JG, April 1995. See also *Surely You're Joking.*
9. *Proceedings of the International Conference on the Theory of Gravitation,* Gauthier-Villars, Paris, 1964.
10. *Surely You're Joking.*
11. Mehra. The comment that 'no one else knew' refers, of course, to how Feynman felt at the time he made his discovery; it was only later, as we discuss in the main text, that he found out that Gell-Mann, and Sudarshan and Marshak, had got there independently, and that didn't diminish his euphoria one bit.

12. See Mehra.
13. *Surely You're Joking.*
14. Mehra.
15. Mehra.
16. It was published in the February 1960 issue of *Engineering and Science*; see also *No Ordinary Genius.*
17. See, for example, Ed Regis, *Nano!* (Bantam Press, London, 1995).
18. Told by McLellan to Gleick.
19. Feynman, note 16.

— 9 —

Fame and (some) fortune

As Feynman entered the 1960s, he was secure in both his personal and his professional lives. He was about to be married, he had made the decision never to leave Caltech, and in the autumn of 1959 he had been appointed Richard Chace Tolman Professor of Theoretical Physics, bringing his salary in 1960 above the $20,000 mark and making him the highest paid member of the faculty. But he was as yet a well-known figure only in the world of physics.[1] By now in his early forties, even Feynman himself may have suspected that his great achievements in theoretical physics all lay behind him, although he continued to beaver away on his investigations of gravity, trying to find a way to a quantum mechanical description of gravitational phenomena, linking gravity and quantum physics in the way that Maxwell had linked electricity and magnetism. He never succeeded in that objective. But his career was about to take an unexpected turn that would lead to much more fame than Feynman can ever have anticipated; and as we shall see in Chapter 10, by the end of the decade, even in his fifties Feynman would make one last great contribution to theoretical physics.

In spite of its success as a world centre for research, at the beginning of the 1960s physics at Caltech had a problem. Undergraduates were still being taught courses along the lines laid down in the 1940s, learning a great deal of classical physics in their first two years, but only coming on to the excitement of topics like relativity, quantum theory and atomic physics in their third year of study, by which time their brains had been numbed by the dullness of the first two years.

The person who started the move to drag physics teaching at Caltech into the second half of the twentieth century was Matthew

Sands, the physicist friend of Feynman who had acted as Gweneth's sponsor when she had applied for her visa. Sands persuaded an initially reluctant Robert Bacher, head of the physics division, that something needed to be done, and Bacher obtained funds from the Ford Foundation towards the cost of a complete overhaul of the introductory physics course. Bacher brought Robert Leighton, a more traditionally minded physicist, on board to act as a counterbalance to some of Sands' more extreme enthusiasms, while the experimenter Victor Neher set to work devising the practical side of the lab work for the new course.

The collaboration between Sands and Leighton did not proceed smoothly in the early months of 1960. Leighton wanted a traditional course; Sands, constantly seeking advice from Feynman, wanted something new and fresh. 'We could not seem to converge on a solution', Sands later told Mehra; but 'one day I had the brilliant inspiration of saying, "Look, why don't we get Feynman to give the lectures and let him make the final decision on the contents?"'

No other great physicist had ever taught freshman physics (at least, not after achieving the status of being a great physicist), and Feynman was intrigued by the challenge, and the opportunity to set out his way of thinking about the world for a wider audience. Leighton was wary, but carried along by the enthusiasm of Sands and Neher. So it was that Feynman began what was supposed to be a one-year series of lectures on introductory physics in the autumn of 1961. In the end, the course spanned two academic years, from September 1961 to May 1963. The deal was that he would give the course once, and once only.

Aware that this was going to be a special event, from the outset Caltech took care to ensure that the lectures were preserved for posterity. Everything was recorded, and Leighton and Sands took on the job of converting the recordings (a total of more than a million spoken words)[2] first into written notes and then into book form (a task which Leighton estimated involved from 10 to 20 hours of work per lecture).[3] Feynman gave two lectures a week, and devoted himself full time to their preparation and presentation, planning how to structure his presentation and get the story across. But although he thought everything through in advance, he had no formal notes when it came time to talk, just a single sheet of paper with key words written on it to remind him of the flow of the presentation.

What made Feynman a great teacher, according to David Goodstein,[4] was that 'for Feynman, the lecture hall was a theater, and the lecturer a performer, responsible for providing drama and fireworks as well as facts and figures. This was true regardless of his audience, whether he was talking to undergraduates or graduate students, to his colleagues or the general public.' Goodstein stresses the amount of preparation that went into all of Feynman's lectures down the years, so that although he was certainly capable of talking spontaneously on almost any aspect of physics, and did indeed include off-the-cuff remarks in his lectures, the whole structure of the talk (including some of the apparent ad libs and joking asides) was carefully planned in advance. 'He didn't need very many notes – I know from his lecture notes, that he didn't need very many notes to remind himself of what he wanted to say. But he knew in great detail what he wanted to say.'[5]

Feynman's famous undergraduate lectures lived up to this ideal; they were like shows, entertainments with a beginning, a middle and an end. Each lecture was self-contained, but ended with a summary of key points that the students were supposed to carry away with them for future reference. Anybody who wants to can now get a flavour of what it was like to be present at these lectures, because in 1995 six of them were presented in a package combining a book and copies of the original audio recordings of Feynman giving the lectures.[6] There is no better way to get a feel for the physics of atoms and molecules, quantum theory, energy, gravity and the relationship of physics to other sciences than by listening to these recordings and reading the book that goes with them. For, of course, Feynman being Feynman, the lectures were not just an introduction to physics for freshmen. They represent a guide to physics as he understood it, the way different pieces fit together, the way to think about things, a philosophy of problem solving.

The originally intended basic course in physics occupies the first two volumes of the published version of the *Lectures*. At the end of the course, in May 1963, Feynman gave himself the challenge of presenting advanced quantum mechanics to a sophomore audience. Together with a republishing of two introductory chapters from volume one, and some additional material developed in 1964, these final lectures formed the basis of volume three of the *Lectures*.

The whole package made a huge impact on physics, and physicists, around the world, although not quite in the way that Feynman and his colleagues had originally intended. Feynman's aim, as he states in the preface to the books, was 'to address them to the most intelligent in the class' while also providing 'at least a central core or backbone of material' for the less able students. It is generally accepted that the lectures failed the less able students, at least partly, and Feynman's course did not become the basis of the formal teaching of physics undergraduates at Caltech (or, as far as we are aware, anywhere). By all accounts, the first-year lectures went well. 'I think', said Feynman in his preface, 'that things worked out – so far as the physics is concerned – quite satisfactorily in the first year.' But while Feynman goes on to comment (referring particularly to the lectures on quantum physics), 'I don't think I did very well by the students', Sands says in his foreword to volume three, 'I believe that the experiment was a success.'

One of us (JG) is well placed to explain this seeming dichotomy. I studied physics (at the University of Sussex) from 1963 to 1966, and read the famous 'red books' as they were published, from 1963 to 1965, alongside the formal coursework for my degree. The Feynman *Lectures* came across as a breathtaking insight into how physics really works, a brilliant counterpart to the formal coursework. For anyone who loved physics (not necessarily, as my example shows, the brightest students, but the ones who really cared about the subject), they provided a goldmine of information and opportunities to go beyond the formal teaching. Anyone with sufficient motivation could indeed learn physics, including quantum physics, from the books; but they work best if you already know some of the story.

Feynman's 'magnificent achievement', in the words of Goodstein,[7] 'was nothing less than to see all of physics with fresh new eyes. Feynman was more than merely a great teacher. His lasting monument is that he was a great teacher of teachers.' And Feynman himself told Goodstein that his most important contribution to physics, in the long run, would not be seen as QED or his other theoretical work, but the Feynman *Lectures*.[8] The point he was surely making is that scientific theories may come and go, being superseded by better theories, but the scientific method, the pleasure of finding things out that he describes so lovingly in these books, is the bedrock upon which all of science is built.

The *Lectures* themselves certainly carry you to the heights – the 'easy pieces' are chosen because they are indeed easy, and should not be regarded as entirely representative. But where else would you get, for example, a complete special lecture on the Principle of Least Action, an almost verbatim record of a great physicist describing one of his own great loves in physics? Before the end of volume two, Feynman gives his readers a summary, involving just nine equations and taking up less than half a printed page of space, which contains all of classical physics, from Newton to Einstein via Maxwell. And – and this is the point – by this point the reader knows that this half-page really does sum up all of classical physics, and can share Feynman's joy at the simplicity of its presentation.

As Feynman himself said in his preface:

> There isn't any solution to this problem of education other than to realise that the best teaching can be done only when there is a direct individual relationship between a student and a good teacher – a situation in which the student discusses the ideas, thinks about things, and talks about things. But in our modern times we have so many students to teach that we have to try to find some substitute for the ideal. Perhaps my lectures can make some contribution. Perhaps in some small place where there are individual teachers and students, they may get some inspiration or some ideas from the lectures. Perhaps they will have fun thinking them through – or going on to develop some of the ideas further.

That is exactly what has happened, to thousands of students and teachers of physics. The Feynman *Lectures* have indeed acted as an inspiration, a source of ideas and a basis for discussion. They have never been out of print in the past three decades, and even provided an inspiration for Feynman himself. Dissatisfied with the lectures on quantum mechanics, he wrote in the preface, 'maybe I'll have a chance to do it again someday. Then I'll do it right.' He did, 20 years later, in the lectures that formed the basis of his book *QED*, probably his most successful attempt at making supposedly 'difficult' ideas in fundamental physics accessible to a wide audience. Not just accessible, but entertaining – Freeman Dyson wrote, in *From Eros to Gaia*, that 'Dick Feynman was a great communicator. I never saw him give a lecture that did not make the audience laugh.'

By the time of the QED lectures, though, Feynman would be extremely well known as a lecturer, author and publicly visible man of science. He was certainly getting plenty of practice at lecturing. During the academic year 1962–63, as well as giving (twice a week) the lectures that would become volumes two and three of the 'red books', every Monday morning in term time Feynman gave a series of postgraduate lectures on gravity, 27 of them in all, summing up his research on the subject.[9] And for light relief, every Wednesday he was off to the Hughes Research Laboratories in Malibu to give his regular informal lecture there. As if that weren't enough to keep him busy, remember that Carl was born in 1962, so there must have been a few sleepless nights during the 1962–63 academic year.

Hot on the heels of the undergraduate lectures at Caltech, in 1964 he accepted an invitation to give a series of lectures at Cornell (in the annual series of Messenger Lectures), choosing as his subject 'The Character of Physical Law'. The lectures were recorded by the BBC and broadcast on TV as well as being turned into a book of the same title. Among other things, Feynman looked at gravity as the archetypal example of a physical law, the relation of mathematics to physics, quantum theory and the distinction between past and future. The book, aimed at a non-scientific audience, helped to establish Feynman in the eye of a broader public as a kind of homespun philosopher of science (much though he would have abhorred being called a philosopher) who had clear and important things to say about the whole basis of the scientific pursuit of knowledge. The heart of that philosophy is summed up in the last of these Messenger Lectures, in ringing tones that not only set out the stall of physics but also, calculatedly, pull the rug from under mysticism of all kinds and speak up for rationalism:

> In general we look for a new law by the following process. First we guess it. Then we compute the consequences of the guess to see what would be implied if this law that we guessed is right. Then we compare the result of the computation to nature, with experiment or experience, compare it directly with observation, to see if it works.

> *If it disagrees with experiment it is wrong.** In that simple statement is the key to science. It does not make any difference how beautiful your guess is. It does not make any difference how smart you are, who made the guess, or what his name is – if it disagrees with experiment it is wrong.

That sentence, *if it disagrees with experiment it is wrong*, ought to be engraved in large letters on the wall of every science department in the world. Less than a year after Feynman had spoken those words at Cornell, on 21 October 1965 the Nobel Committee acknowledged the paramount example of the best agreement that had ever been found between experiment and theory, when they gave the 1965 Nobel Prize for Physics to Feynman, Schwinger and Tomonaga for their work on QED.

In *Surely You're Joking*, Feynman recounts that he seriously considered refusing to accept the prize. The Nobel Committee does not check out in advance whether somebody actually wants their award, they just announce it to the world, at a convenient time of day in Stockholm – which means the middle of the night in California. The first Feynman heard of it was when he was woken early in the morning, some time before 4 a.m., by the telephone, as people called to offer congratulations and reporters asked for comments. Having taken the phone off the hook and sat down in his study to think about the implications, he wondered if it was worth going through the hoopla and publicity of accepting the prize; he knew that in many cases Nobel Prize-winning physicists had become figureheads, involved in administration, giving guest lectures here and there, and sure of a job for life, but no longer active in scientific research. When he put the phone back on the hook and let it ring again, one of the first calls was from *Time* magazine. Feynman asked the reporter if there was some way not to accept the prize; the reporter pointed out that it would be a much bigger and more sensational news story if he turned it down than if he accepted.[10]

Nobody else had any doubts about the wisdom of accepting the prize. The Caltech students draped the administration building, Throope Hall, with a banner reading 'WIN BIG, RPF', and his

*Our emphasis.

colleagues were delighted. The interviews were a chore, though, with many reporters asking Feynman to explain his award-winning work for them in one sentence. He later said that he regretted not taking the advice of the representative from *Time*, who had suggested he told them that if it could be described in a sentence it wouldn't be worth the Nobel Prize.[11] The days following the news were full of the kind of distractions that might well stop any scientist ever getting down to serious work again.

It wasn't just the prospect of being regarded as – or, worse, actually being – a has-been once the award was conferred. For all his efforts to appear brashly unconcerned about formality (whenever he dined at the Athenaeum Club on campus at Caltech, even if he had worn a jacket and tie in to work Feynman always made a point of walking over in shirtsleeves, and selecting the most garish of the neckties that the Club held in reserve for diners who had 'forgotten' theirs), Feynman was actually worried about how to cope with all the pomp of the occasion of the award ceremony in Stockholm. He would even, after all his father had told him, have to dress up in a kind of uniform, in white tie and tails like something out of a Fred Astaire movie.

In the end, though, Feynman managed to have fun in Stockholm. He particularly enjoyed a ceremony at a party organized by students, in which the students gave each Nobel Prize-winner the 'Order of the Frog', which involved the recipient making an appropriate frog noise. It just happened that Feynman knew exactly how to make a frog noise, having seen his father's copy of the Aristophanes play *The Frogs* when he was a boy. In the play, Aristophanes describes frogs as going 'brekebek, brekebek'; young Richard had thought this was silly, but tried it out and found that it really did sound like a frog. The never-forgotten skill came in useful in Stockholm.

The Feynmans got to dine with royalty in Stockholm (a mixed experience, given Richard's feelings about uniforms and authority) and were treated like royalty themselves for a few days, chauffeured around to see the sights. The award ceremony itself took place on 11 December 1965, and one of the few duties required of the recipients was to give a lecture about their work. Feynman chose not to describe QED itself, leaving that for Schwinger and Tomonaga; instead, he described his path to quantum electrodynamics, the sequence of ideas that had led him to his great work. For later generations, this was far and away the best thing that happened at the 1965 Nobel ceremonies, providing us with an inside view of the development of Feynman's version of QED, from early ideas about direct action and advanced potentials right through to Feynman's views on the best way to make progress in theoretical physics, by guessing solutions to problems and comparing the guesses with experiments.[12]

'All in all,' Feynman decided, 'I enjoyed the visit to Sweden, in the end.'[13] Not least, of course, because of the fun Gweneth had on the trip, and also because it provided him with a fund of new anecdotes. Before returning to the United States, for example, he went to Switzerland to give a talk at CERN, the European centre for particle physics research. He wore the suit he had worn to have dinner with the King of Sweden, and began his lecture, tongue in cheek, by telling his audience of physicists how he had been changed by the award, and had decided that he rather liked wearing the suit to give a lecture. The audience erupted into jeers and catcalls; Feynman stripped off his jacket and tie, grinning his famous grin, and after the laughter had died down continued his lecture in shirtsleeves, as usual. It was CERN, he liked to say, that had straightened him out after Stockholm; the CERN audience 'undid everything that they had done in Sweden'.

To prove that he was still the same Feynman, Dick accepted a bet from Viktor Weisskopf, then the Director of CERN, during that visit. Feynman agreed in writing that he would pay $10 to Weisskopf if at any time during the next 10 years 'the said Mr Feynman has held a "responsible position"'. Conversely, if Feynman had not held a responsible position in that time, Weisskopf would pay him $10. 'The term "responsible position" shall be taken to signify a position which, by reason of its nature, compels the holder to issue instructions to other persons to carry out certain acts, notwithstanding the fact that the holder has no understanding whatsoever of that which he is instructing the aforesaid persons to accomplish.' Weisskopf was obviously drawing on his own experience as an administrator in drafting that definition; but he lost his bet. Feynman collected his $10 from Weisskopf in 1976, and never held a responsible position in his life.[14]

As Feynman later observed on many occasions, the lasting pleasure he had found in the Nobel Prize was to discover how many people loved him. 'He had', says Mehra, 'found a genuine thrill in the messages of congratulations, expressing love and affection and admiration, many of them from school children and students.' But there remained one deep sadness associated with the prize. Melville hadn't lived to see it. Many years later, during one of their drumming and storytelling sessions, Ralph Leighton asked Feynman, 'If you could, which figure from the past would you most like to bring back and talk to?', imagining that Feynman might pick Newton, or Galileo, or some other scientific figure. Richard replied, 'I'd like to bring back my father, so I could tell him I won the Nobel Prize.'[15]

As well as fame, the Nobel Prize brought with it some fortune – a one-third share in $55,000. The Feynmans used it to buy a beach house in Mexico, at Playa de la Mision in Baja California. Gweneth was as adventurous as Dick, and both of them loved travelling, often backpacking and sleeping rough; but a slightly more civilized holiday home was no bad thing with Carl, now three years old, to consider.

Richard's relationship with Carl echoed, in many ways, his own relationship with Melville. He would explain things about the way the world worked, expressing his love of science and sharing it with his son, without pushing him in any particular direction. Carl was interested, and responded in the way Richard must have hoped when he made up stories involving scientific insights into the nature of the world. There came a time when, to Feynman's horror, Carl decided to study philosophy as an undergraduate in college. But it all worked out in the end, because he ended up in computer science, which made his father much happier.

When Feynman tried the same approach on Michelle, though, it didn't work. She didn't want him to make up stories full of interesting insights about the world, but to read familiar stories from a book over and over again. So the Feynman method of encouraging children to become scientists doesn't always work; it depends on the personality of the child.[16]

But although his home life was happy and comfortable, his fame was assured and he had no financial worries, in the years just after he received the Nobel Prize Feynman began to fear that he really was burned out. He tried to cut himself off from all of the distractions associated with his growing prestige; as he had promised himself long ago, when sharing the graduation ceremony at Princeton for the degree he had worked for with honorary graduates who had done nothing to earn their degrees, when offers of honorary degrees started to come his way he turned them all down. The first offer of this kind came at the beginning of 1967, and was politely declined, as were all later offers. The offer came from the University of Chicago, and Feynman replied:

> Yours is the first honorary degree that I have been offered, and I thank you for considering me for such an honor. However, I remember the work I did to get a real degree at Princeton and the guys on the same platform receiving honorary degrees without work – and felt that an 'honorary degree' was a debasement of the idea . . . It is like giving an 'honorary electrician's license'. I swore then that if by chance I was ever offered one I would not accept it. Now at last (twenty-five years later) you have given me a chance to carry out my vow.[17]

Feynman also turned down invitations to visit research centres around the world to give guest lectures, unless the invitation came from somewhere he liked to visit, such as Brazil or Japan. He was trying to keep himself free from time-consuming commitments, in the hope of making more progress in physics, even though he was now approaching his fiftieth birthday and had been encumbered with the status of a Nobel laureate.

But this didn't stop Feynman from having fun. The weekends in Las Vegas might be a thing of the past, but there were plenty of avenues to explore in Pasadena itself. In the 1960s, Feynman's big interest, outside science and his family, was art, and this led him into one of his most famous escapades.

The interest in art came about through Feynman's friendship with Jirayr ('Jerry') Zorthian, an artist he met at a party in the late 1950s. Although both Jerry and Dick were extrovert party animals, at first their friendship was based on an attraction of opposites, Jerry intrigued by a chance to get to know how scientists dispassionately view the world, while Dick was fascinated by what he saw as the excessive freedom of the artist, working with so few rules that, it seemed, anything went. He had the attitude, 'What is this contemporary art? A child can do better', to which Jerry responded by giving him a crayon and asking him to do better himself.[18] The arguments came to a head when Feynman suggested that they resolve the situation by each learning about the other's craft. On alternate Sundays, he would give Jerry a lesson in science, while Jerry gave him lessons in art on the Sundays in between.[19]

Feynman became an accomplished amateur artist, developing from his lessons with Zorthian to more formal tuition, and eventually having an exhibition all of his own. This led to a hilarious encounter. One of the drawings in the exhibition had started out as an exercise in shading, and was a portrait of a nude model lit from below and to one side. For the exhibition, he whimsically gave it the title, 'Madame Curie Observing the Radiations from Radium'. An art-lover at the showing came up and asked if he drew from photographs, or using live models. Always, Feynman replied, from live models. Then how, came back the puzzled inquiry, did you persuade Madame Curie to pose for you?[20]

Zorthian's attempts to learn science were not so successful, and the experiment petered out almost immediately. That gave Jerry and Dick scope for a new on-going argument: whether Jerry was a better teacher than Dick, or Dick was a better student than Jerry.

Feynman also sold some of his pictures (he signed them with a pseudonym, 'Ofey'), and enjoyed the whole new experience of being part of the art scene. Among the many new friends he made through his drawings was Gianonni, the owner of a topless bar in Pasadena. Gianonni's was only about a mile and a half from Feynman's home, and Richard found it a convenient place to go and sit, drinking 7-Up and quietly working on some physics problem, or doing a little sketching, in one of the booths at the back. Gweneth was unfazed by this, regarding the bar as Feynman's equivalent of one of the traditional gentlemen's clubs back in England. The one drawback about the bar, as far as Feynman was concerned, was the pictures on the walls, crudely titillating efforts in garish colours. He offered Gianonni one of his own nudes instead; the owner of the bar was so pleased that he not only put the picture up on the wall but gave instructions for Feynman to have free 7-Up whenever he came into the place.

Eventually, towards the end of the 1960s, the bar was raided by the police, and an attempt was made to close it down. There was a big court case, and Gianonni asked all his loyal customers to testify on his behalf, that there was nothing lewd and obscene going on in the bar. Of course, they all made their excuses – except one. So Feynman testified in court that he was a regular at the bar, that many respectable pillars of the community from all walks of life were also regulars there, and that nothing went on that could be regarded as an offence to the community. Hardly surprisingly, the testimony made headlines – 'Caltech's Feynman Tells Lewd Case Jury He Watched Girls While Doing Equations', the local paper gleefully reported on 8 November 1969. Gianonni lost the case, but the bar stayed open pending appeal, and Feynman continued to get his free 7-Ups. It was just as well that Feynman still had a convenient place to do his thinking, because by now, at the end of the 1960s and past his own fiftieth birthday, he was well into his stride as a physicist again, in the process of making his last great contribution to science.

The recovery from the trough he had experienced after the award of the Nobel Prize had begun in 1967, when he visited the University of Chicago. The slump in Feynman's creative physics activity had really begun in 1961, when he had finished most of his work on gravity and made the commitment to two years' concentrating on the *Lectures*. It had been the longest more-or-less fallow period of his life in physics; but it is somehow an appropriate part of the Feynman legend that what put him back on the track to real scientific creativity was not an encounter with a new idea in physics, but an encounter with a molecular biologist, James Watson.

Feynman had got to know Watson during the sabbatical year that Dick had spent as a 'graduate student' in biology. He had an opportunity to renew the acquaintance when he visited Chicago early in 1967, and when they met Watson gave Feynman a copy of the typescript of what was to become his famous book *The Double Helix*, about his discovery, together with Francis Crick, of the structure of DNA.[21] Feynman read the book straight through, the same day. He had been accompanied on that trip by David Goodstein, then a young physicist just completing his PhD at Caltech, and late that night Feynman collared Goodstein and told him that he had to read Watson's book – immediately. Goodstein did as he was told, reading through the night while Feynman paced up and down, or sat doodling on a pad of paper. Some time towards dawn, Goodstein looked up and commented to Feynman that the surprising thing was that Watson had been involved in making such a fundamental advance in science, and yet he had been completely out of touch with what everybody else in his field was doing.

Feynman held up the pad he had been doodling on. In the middle, surrounded by all kinds of scribble, was one word, in capitals: DISREGARD. That, he told Goodstein, was the whole point. That was what he had forgotten, and why he had been making so little progress. The way for researchers like himself and Watson to make a breakthrough was to be ignorant of what everybody else was doing, and plough their own furrow.[22] In a letter to Watson that is preserved in the Caltech archive, Feynman wrote 'you are describing how science *is* done. I know, for I have had the same beautiful and frightening experience.'

In fact, what Watson was describing was how science is done not by ordinary scientists, but by those rare individuals who have the ability to achieve new insights and make major breakthroughs. Watson himself had the 'beautiful and frightening experience' once, and earned a Nobel Prize for his efforts. Such a singular achievement is far beyond the realistic aspirations of the great majority of scientists. Even Dirac reached the pinnacle only twice, first with his version of quantum mechanics, then with the equation of the electron. Yet by the time Feynman wrote those words, he had already had that beautiful and frightening experience three times, in his work on QED, on superfluidity and (best of all, in his own eyes) on the weak interaction. And now, he was about to experience it again, as he stopped trying to keep up with the scientific literature or compete with other theorists at their own game, and went back to his roots, comparing experiment with theory, making guesses that were all his own, and coming up with an insight that would give an enormous impetus to the development of particle physics in the 1970s. He would show that there was, indeed, life as a theoretical physicist of the top rank beyond the Nobel Prize.

Notes

1. But still, *very* well known in the world of physics. Interviewed by JG in October 1995, Norman Dombey, who was a graduate student in physics at Caltech in the early 1960s, said that at that time 'Feynman was *the* guy in the subject, and had been since Los Alamos. The only other person on a par was Landau, and Feynman thought that too; he regarded Landau as his Soviet equivalent.' The reference is to Lev Landau, of liquid helium fame.
2. Sands, foreword to volume two of *The Feynman Lectures on Physics*, hereafter referred to as the *Lectures.*
3. Leighton, foreword to volume one of the *Lectures.*
4. See *Most of the Good Stuff.*
5. Interview with JG, April 1995. This is similar to the way in which Feynman honed his anecdotes about his own life to tell the truth, but to tell it in an entertaining way.
6. *Six Easy Pieces* (see Bibliography).
7. See *Most of the Good Stuff.*
8. David Goodstein, interview with JG, April 1995.
9. Thanks to notes taken by two of the graduate students who attended those lectures, the first 16 of them, roughly covering the work up to the point where Feynman ran into a brick wall, were eventually published, in 1995, as *Feynman Lectures on Gravitation.* Their significance today is discussed in Chapter 14.
10. Jagdish Mehra says that exactly the same thing occurred when Dirac was offered the Nobel Prize. He wanted to turn it down, but was advised by Ernest Rutherford that he would get more publicity by refusing the prize than by accepting it. Several times, Dirac commented to Mehra that the prize had been 'a nuisance'.
11. Mehra.
12. Nobel lecture, *Science*, volume 153, page 699, 1966.
13. *Surely You're Joking.*
14. Mehra.
15. Leighton, interview with JG, April 1995.
16. See contributions by Carl and Richard to *No Ordinary Genius.*
17. Letter in Caltech archive; also quoted by Schweber.
18. See Zorthian's contribution to *No Ordinary Genius.*
19. *Surely You're Joking.*
20. *Surely You're Joking.*
21. The best available edition of Watson's famous book is the 'critical edition' edited by Gunther Stent (Weidenfeld & Nicolson, London,

1981). This includes all of Watson's text (originally published in 1968), plus reviews, commentaries and reprints of some of the original scientific papers.

22. David Goodstein, interview with JG, April 1995; see also Gleick.

10

Beyond the Nobel Prize

Albert Einstein was almost unique among the physicists of modern times in making major contributions to fundamental physics in each of three separate decades – the 1900s, the 1910s and the 1920s. Born in 1879, he completed his last important work, involving the application of Bose–Einstein statistics, in the mid-1920s, a couple of years short of his own fiftieth birthday. But his achievement is only 'almost' unique because it has been matched by one other physicist, Richard Feynman, who made major contributions to fundamental physics in the 1940s, 1950s and 1960s. Indeed, Feynman's last great work continued well into the 1970s, and occupied him until only just before his own *sixtieth* birthday. In the words of David Goodstein, 'even among Nobel Prize-winners, he was extraordinary. Long before he won the Nobel Prize, he was a legend in the community of scientists.'[1]

Feynman made his name in the 1940s with his work on QED, providing a theory of one of the four fundamental forces (or interactions) of nature, electromagnetism. In the 1950s, as we have seen, he made a major contribution to developing physicists' understanding of another fundamental force, the weak interaction, and then went on to make a major contribution (only fully appreciated in the 1980s and 1990s) to the understanding of a third force, gravity. His work in the late 1960s and early 1970s provided profound insights into the workings of the fourth force, the strong interaction. Nobody else has made such influential contributions to the investigation of all four of the interactions – even Murray Gell-Mann, for example, made significant contributions only to the study of two of the interactions (the strong and the weak), and he is generally regarded as a remarkable genius.

Gell-Mann, who worked in the office just down the hall from Feynman at Caltech, was closely involved in theoretical investigations of the particle world in the 1950s and 1960s, and helped to bring some sort of order into the chaotic confusion of particles that had been discovered as the new particle accelerators had probed to higher and higher energies. Although he and Feynman had collaborated, in memorable fashion, on one important piece of work concerning the weak interaction, their styles and approaches to physics were so different that it was inevitable that they would largely go their own ways, although it was convenient for each of them to have the other to bounce ideas off on occasion. Was there a rivalry between them which helped to spur each of them on? Norman Dombey, one of Gell-Mann's former students, says that 'I think it spurred Gell-Mann on. He couldn't stand anybody beating him.'[2] If so, he was certainly spurred on to good effect.

Back in the early 1930s, physicists had known of just four fundamental particles, to set alongside the four fundamental interactions. All you needed to explain the properties of everyday atomic matter were the proton, the neutron and the electron, together with the neutrino, which had never been detected directly, but was needed to explain details of beta decay. Then, 'new' particles began to turn up – very short-lived particles, which quickly decayed into the familiar stable particles and intense pulses of electromagnetic radiation, but real none the less, with distinctive properties (such as mass and charge) that could be measured during their brief lives. The first of these particles were found in showers of cosmic rays. Then, after the Second World War, physicists began to build the 'atom smashing' machines in which they could create exotic particles more or less at will.

This work involves using electromagnetic fields to accelerate particles such as electrons and protons to high velocities (a sizeable fraction of the speed of light), and then smash the beams of high-energy particles into either a target of ordinary matter, or into another beam of particles going in the opposite direction. When some of the particles in such a beam are brought to a sudden halt in the resulting collisions, their energy of motion (the kinetic energy) is released, and is available to manufacture other particles, in line with Einstein's equation $E = mc^2$.

It is important to stress that the exotic particles are manufactured out of pure energy. If a fast-moving electron collides with a neutron, say, and produces a shower of particles, this does not mean that those particles were in any sense hidden inside the neutron waiting to be liberated; in such experiments, the combined mass of the particles produced in the collision may be many times more than the mass of the neutron, and all this mass has come from the energy of motion of the colliding particles.

By the end of the 1950s, dozens of different kinds of particles were known that could be produced out of energy in this way, live their brief lives, then decay into a mixture of high-energy photons and ordinary stable particles. How could such a profusion of particles be regarded as in any sense 'fundamental'? How could some sort of order be brought into the chaos?

The first step was to group the particles according to their common properties. There are two key criteria. Particles which are affected by the strong force (such as protons and neutrons) are called baryons. Particles which are not affected by the strong force (such as electrons) are called leptons. Baryons and leptons are all fermions. In each case, there are force-carrying bosons (such as the photon), with the ones that carry the strong force generally referred to by the overall name of mesons. And mesons and baryons are together often referred to as hadrons. The embarrassing proliferation of particles in the 1950s chiefly involved hadrons, with both new baryons and new mesons turning up by the handful.

In 1961, Gell-Mann and the Israeli physicist Yuval Ne'eman (then working at the University of London, in England) independently hit upon a way of arranging hadrons in accordance with their properties (mass, charge and so on) in a pattern that Gell-Mann dubbed 'the eightfold way', because it grouped the particles in octets. The approach was very similar to the way Dmitri Mendeleyev had grouped the chemical elements into the pattern that we now call the Periodic Table, back in the 1860s. Just as Mendeleyev's arrangement of chemical elements only worked if certain gaps were left in his table, corresponding to elements that had not yet been discovered, so the eightfold way classification only worked if certain gaps were left in some of the octets, corresponding to particles that had not yet been discovered. And, just as Mendeleyev was triumphantly proved correct when new chemical elements were found with exactly the properties required to slot them into the gaps in his table, so Gell-Mann and Ne'eman were triumphantly proved correct when new particles were found with exactly the properties required to slot them into the gaps they had left in their classification. For this and his other work on the classification of fundamental particles, Gell-Mann received the 1969 Nobel Prize for Physics; surprisingly, the Nobel Committee overlooked Ne'eman.

The order in the Periodic Table of the Elements is explained, of course, because atoms are not indivisible. The properties of atoms are determined by the number and nature of the particles they are made of – the electrons, protons and neutrons. It was natural to guess that the order in the eightfold way classification might be explained if hadrons were also composed of different arrangements of some sort of truly fundamental particles. But physicists were so used to thinking of protons and neutrons, in particular, as indivisible fundamental entities that it took a long time for the idea that they might be composite entities to become accepted. It was in making this concept (of protons, neutrons and other baryons being composite particles) acceptable that Feynman made his next great contribution to physics. But he was not the first on the trail, because in the early 1960s he was finishing up his work on gravity and becoming deeply immersed in his undergraduate lectures.

The first tentative steps towards the idea of a deeper layer of particles within the hadrons was made in 1962 by Ne'eman (then working for the Israel Atomic Energy Commission) and his colleague Haim Goldberg-Ophir. They wrote a paper suggesting that baryons might each be made up of three more fundamental particles, and sent it to the journal *Il Nuovo Cimento*, where it was mislaid for a time, but was eventually published in January 1963. The paper attracted little attention, partly because the eightfold way itself had not yet been fully accepted, but also, as Ne'eman has acknowledged, 'because it did not go far enough. The authors had developed the mathematics resulting from the eightfold way, but they had not yet decided whether to regard the fundamental components as proper particles or as abstract fields that did not materialize as particles.'[3]

One person who had no such inhibitions was George Zweig, a

PhD student at Caltech. Zweig had been born in Moscow in 1937, but moved to the United States as a baby and obtained a BSc in mathematics from the University of Michigan in 1959. He started his research career at Caltech as an experimental particle physicist, but after three years struggling with a recalcitrant experiment on an accelerator called the Bevatron, he decided that experiment was not his forte and turned to theoretical physics, under the nominal guidance of Richard Feynman but actually working largely on his own. Zweig was immediately taken with the beauty and simplicity of the eightfold way, and quickly realized that the pattern of octets could be explained if mesons and baryons were composed, respectively, of pairs and triplets of fundamental entities, which he called 'aces'. Zweig regarded these, from the outset, as real particles, not 'abstract fields', and he was unfazed by the fact that in order to make the scheme work each of his aces would have to have a fraction of the charge on the electron – either $\frac{2}{3}$ or $\frac{1}{3}$, in units where the electron's charge is 1.

Although Zweig wrote up his ideas for publication, they met with such a violent response that the papers were never formally published in their original form. In 1963, on a short term visit to CERN, Zweig prepared two papers which were circulated in the form of CERN 'preprints', but as he later recalled:

> Getting the CERN report published in the form that I wanted was so difficult that I finally gave up trying. When the physics department of a leading university was considering an appointment for me, their senior theorist, one of the most respected spokesmen for all of theoretical physics, blocked the appointment at a faculty meeting by passionately arguing that the ace model was the work of a 'charlatan'.[4]

As if this were not bad enough, Zweig's work was soon to be overshadowed by Gell-Mann, who had hit on the same idea, completely independently, back at Caltech. But Gell-Mann was much more cautious, and trod a path almost exactly halfway between the confident espousal of aces as real by Zweig, and the dismissal of the 'fundamental components' as 'abstract fields' by Ne'eman and Goldberg-Ophir. Like Zweig, he gave the fundamental entities a name ('quarks'); but like the Israeli team he expressed reservations about their reality. In a paper that was published in *Physics Letters* in 1964, Gell-Mann said:

> It is fun to speculate about the way quarks would behave if they were physical particles of finite mass (instead of purely mathematical entities as they would be in the limit of infinite mass) . . . a search for stable quarks of charge $-\frac{1}{3}$ or $+\frac{2}{3}$ and/or stable diquarks of charge $-\frac{2}{3}$ or $+\frac{1}{3}$ or $+\frac{4}{3}$ at the highest energy accelerators would help to reassure us of the non-existence of real quarks![5]

This is an astonishingly oblique way of presenting a great new idea in physics, and one which Gell-Mann lived to regret. With hindsight, it is probably unfortunate that Zweig was away from Caltech when he developed the theory of aces. Back in Pasadena, he would have had the chance to discuss the idea with Feynman, and almost certainly the Caltech authorities would have urged a joint publication with Gell-Mann, just as Feynman and Gell-Mann had been forced into a fruitful shotgun marriage with their work on the weak interaction. A joint paper by Gell-Mann and Zweig, less overtly cautious than Gell-Mann's paper but not triggering the same knee-jerk reaction as Zweig's preprints, and endorsed by Feynman, might well have made more of a splash in 1964 than either of their solo efforts.

As it was, it took a long time for physicists to become convinced that anything was going on inside the hadrons. When physicists did become convinced of the reality of these entities inside the baryons, it was Gell-Mann's name, not Zweig's, that stuck. According to Gell-Mann,[6] he chose the name as a made-up nonsense word, meaning it to rhyme with 'pork', and only later realized the relationship to the passage in James Joyce's *Finnegans Wake*, referring to 'three quarks for Muster Mark', which suggests a pronunciation to rhyme with 'bark'. But since Gell-Mann had previously read *Finnegans Wake* several times, the association may have been there in his subconscious all along. Either way, both pronunciations are used today.

The importance of all this confusion is that this really is the way things were in the middle to late 1960s – confused. Most people regarded the quark model as a wild idea; even Gell-Mann seemed to be at best half-hearted about it, and the one person who had vigorously promoted it found his career prospects severely damaged as a result. Gell-Mann continued to develop the idea (with fewer reservations), but as experiments at high-energy accelerators never did find any evidence of free particles with fractional charge, many physicists found it hard to believe in the reality of quarks.

Gell-Mann, by now, was nearing the end of his career as a great original thinker. He had been born in 1929, did his best work between about 1954 and 1964 (between the ages of 25 and 35), was appointed R. A. Millikan Professor of Theoretical Physics at Caltech in 1967, and received the Nobel Prize in 1969, settling down as a wise older member of the science community but making only relatively minor contributions to fundamental physics after he entered his forties. This is very much the pattern associated with an ordinary genius, and it might have been natural to expect the next leap forward to come from a member of the younger generation, like Zweig. In fact, it was made by a man who was 11 years older than Gell-Mann, and who was just entering his fifties.

It is a sign of how little faith the physics community had in the quark model that in 1969 the citation for Gell-Mann's Nobel Prize rather pointedly avoided any reference to the idea, mentioning instead his earlier work on the classification of elementary particles and their interactions – in other words, the eightfold way and the theory of the weak interaction.[7]

Within a year of his encounter with James Watson in Chicago, where he relearned the lesson that the way to make progress was to disregard what others were doing and start from first principles, Feynman was getting to grips with the theory of what happens in collisions between hadrons – for example, when a proton collides, at very high speed (that is, high energy) with another proton (or an antiproton). This was in 1968, the year Feynman turned 50, and the year in which Michelle joined the family. He developed a model – a way of describing what went on in such collisions – by regarding each hadron as a cloud of point-like particles. He was deliberately agnostic about the nature of those internal constituents – they might be quarks, or they might not. As ever, Feynman was solving the *general* problem, for any number of particles with whatever individual properties they might happen to have, not looking at any particular special case; even at this late stage of his career, the work had all the hallmarks of a classic piece of Feynman research, right down to the mathematical toolkit he used to tackle the problem.

This wasn't out of sheer stubbornness. Feynman's insistence on always trying to solve problems in the most general form, with as few preconceptions as possible, was part of his philosophy, a way of ensuring that you, the researcher, stayed honest in the game of developing theoretical models to explain, or (better) predict what was happening in experiments. Giving the Caltech commencement address in 1974, smack in the middle of his last great burst of creativity as a physicist, he would tell his audience of aspiring scientists about the importance of absolute integrity in science, that 'the first principle is that you must not fool yourself – and you are the easiest person to fool. So you have to be very careful about that. After you've not fooled yourself, it's easy not to fool other scientists. You just have to be honest in a conventional way after that.'[8] This is why he made no presumptions about the nature of the internal constituents of the hadrons, and dubbed those constituents 'partons', a rather ugly word indicating that they were parts of a hadron, but one which carried no load of expectation or preconception about the nature (or even the number) of the particles.

You might think of a swarm of such partons inside a hadron as like a swarm of bees, moving around in a roughly spherical volume of space. But when the hadron is moving at close to the speed of light, as Feynman realized, strange relativistic effects come into play. The sphere is squashed in the direction of its flight, as seen by somebody at rest in the laboratory, to become a highly flattened pancake. For example, a sphere travelling at 0.999957 times the speed of light (this has been achieved in this kind of experiment) shrinks to $\frac{1}{108}$ times its rest thickness along the line of sight, but stays the same diameter at right angles to the line of flight, becoming a pancake 108 times wider than it is thick. When two such pancakes smash into one another broadside on, according to the parton model most of the partons inside would pass right past one another and off into the sunset. But, just occasionally, two of the partons themselves would collide, slowing dramatically and releasing energy in the form of a flood of 'new' particles. This was the basis of Feynman's model, in which the probability of a collision between two hadrons can be considered as the sum of the probabilities of collisions between individual partons – a mathematical formalism echoing the sum over histories idea.

Feynman worked all this out in the first half of 1968, and developed the insight into a mathematical model, containing the basis of lots of predictions that could be compared with experiment. For, of course, *if it disagrees with experiment it is wrong*. Just at that time, a new particle accelerator had been built at Stanford University, in northern California. It was called the Stanford Linear Accelerator Center, or SLAC, and used a straight, 2-mile-long tube to fire a beam of electrons at a target, where they collided with stationary protons, producing debris in the form of particles streaming out from the point of collision. By monitoring the showers of particles produced in this way, the researchers hoped to find out what protons were like inside. Such an experiment was a few steps short of colliding protons, but since electrons can be regarded as point-like particles, the hope was that by scattering the electrons off protons the experiment would reveal any structure inside the protons, in the same way that scattering particles at much lower energies off atoms had revealed, decades before, the existence of the nucleus inside the atom.

These experiments were being carried out jointly by a team of researchers from MIT and SLAC, led by Jerome Friedman, Henry Kendall and Richard Taylor (similar investigations were carried out at about the same time by researchers at the Deutsche Elektron SYnchrotron Anstalt, or DESY, in Germany). The early results from these experiments were being interpreted by a Stanford theorist, James Bjorken, who has described Feynman's arrival on the scene, and his influence on the development of particle physics at the end of the 1960s, in his contribution to *Most of the Good Stuff*.

Bjorken obtained his PhD from Stanford in 1959, and recalls how, like many other physicists, as a graduate student in the late 1950s he learned quantum electrodynamics the old-fashioned way, ploughing through what was essentially a 1930s style course with 'a seemingly endless, gloomy, turgid mass of field-quantization formalism'. But then came a revelation – 'when Feynman diagrams arrived, it was the sun breaking through the clouds, complete with rainbow and pot of gold. Brilliant! Physical and profound! It was instant conversion to discipleship.'

Something very similar happened with partons. Bjorken joined Stanford University just after completing his PhD, and soon became a tenured member of the faculty there. By 1967, he was a full professor at SLAC. At the time Feynman arrived on this particular scene, Bjorken had been developing a theoretical description of what was going on in the electron–proton collisions at SLAC using a highly mathematical formalism known as current algebra, largely developed, as it happens, by Gell-Mann. He had plotted out graphs of what went on at different energies during the collisions, but had no simple physical picture of what was going on. In the summer of 1968, Feynman happened to be visiting his sister, Joan, who was living near SLAC at the time, and in August he went over to SLAC to see what the experimenters were up to. Bjorken was away, but the experimenters and the other theorists showed Feynman the results Bjorken was coming up with, as well as the raw data. The key feature of this work was that the data looked the same – the graphs had the same shape – whatever the energy of the interactions. This is known as scale invariance. Although Bjorken's colleagues at SLAC could not explain to Feynman where Bjorken had come up with this prediction, which matched the experimental results, Feynman realized that it echoed his own work on partons, using the relativistic pancake description of particle interactions.

'It took Feynman only an evening of calculation with his partons to interpret what was going on', says Bjorken. He returned to SLAC just before Feynman was due to leave, and:

> found much excitement within – and beyond – the theory group there. Feynman sought me out and bombarded me with queries. '*Of course* you must know this . . . *Of course* you must know that . . . ,' he kept saying. I knew about some of the things Feynman mentioned; others I didn't know. And there were things I knew at the time but he did not. What I vividly remember was the language he used: it was not unfamiliar, but it was distinctly *different*. It was an easy, seductive language that everyone could understand. It took no time at all for the parton model bandwagon to get rolling.[9]

Everyone could understand. Just as Schwinger had sneered at the way Feynman's version of QED brought computation to the masses, so Gell-Mann sneered at what he called 'Feynman's put-ons', which made particle physics theory accessible even to people who could not handle the complexities of current algebra.

Feynman went back to SLAC in October 1968 to give a talk about his ideas, and the parton model swept through the team there like wildfire. Over the next few years, experiment and theory developed hand in hand, and it gradually became clear that a version of parton theory in which the partons were identified with quarks could best explain the experimental results. But the power of the parton model was that it also allowed for the possibility of other entities, besides quarks, residing inside protons and neutrons. Feynman was convinced from the outset that quarks – if they did exist – could not be isolated particles any more than an electron could be an isolated particle. Electrons, remember, are surrounded by clouds of virtual photons, the carriers of the electromagnetic force; the current picture of the situation inside a proton or a neutron is that the quarks are associated with clouds of 'gluons', the carriers of the strong force, which holds them together. Parton theory automatically took account of this kind of possibility.

The early version of the theory was largely developed by Bjorken and his colleague Emmanuel Paschos at SLAC; the verification of the reality of quarks was acknowledged in 1990 by the award of the Nobel Prize to Friedman, Kendall and Taylor for the experimental side of the work. Feynman would have approved of this recognition that in physics experiment is king. In 1988 he said,[10] 'I am now a confirmed quarkanian!' As Bjorken has put it, referring to Feynman's eventual espousal of quarks, 'it was data that forced the commitment (for both of us)'.[11]

It was well into the 1970s, though, before the blending of quarks and partons was complete, and Feynman himself still had a significant contribution to make to the development of the quark model. As unconcerned as ever about the inglorious rush for priority, he did not hasten to publish these ideas (although he did give several talks on parton theory at scientific gatherings), and his first paper on the subject, written with two of his students, only appeared in the *Physical Review* in 1971, with the cautious comment, 'a quark picture may ultimately pervade the entire field of hadron physics'.[12]

But there was still a major problem with quark theory. If particles existed with a charge one-third or two-thirds of the size of the charge on an electron, why had nobody seen them? Of all the properties they could have, the fractional charge was a distinctive feature that could be observed in very simple experiments. If quarks were real, the only reason that fractional charges were never seen in nature must be because somehow they were kept locked up, or confined, inside hadrons, and could not roam freely about in the world. In that case, you could always ensure that the combined charge on a meson, made of a pair of quarks, added up to round numbers such as zero ($+\frac{1}{3}$ together with $-\frac{1}{3}$) or 1 ($+\frac{2}{3}$ together with $+\frac{1}{3}$), and similarly for baryons suitable triplets of quarks would give, for example, ($+\frac{2}{3}$ together with $-\frac{1}{3}$ and $-\frac{1}{3}$) or ($+\frac{2}{3}$ together with $+\frac{2}{3}$ and $-\frac{1}{3}$).

The picture that emerged was one in which the force that binds quarks together must get *stronger* when the quarks are farther apart. This is both strange and quite natural. In physics, we are used to dealing with forces between two objects that, like gravity or magnetism, are stronger for objects that are closer together. On the other hand, in the everyday world we have a simple example of a force that gets stronger as distance increases. Try stretching an elastic band, and you will literally get a feel for the force between quarks.

Imagine a collision between two quarks which are components of relativistic pancakes travelling in opposite directions. Consider just one of the quarks, happily sitting in a triplet, that receives energy from a head-on collision with a quark in the other pancake and recoils, moving away from its partners. At first, it moves away freely. But the further the quark is going to move, the more energy will be required to drag it apart from its companions. If there is not enough energy available, the quark will snap back into place, like a stretched elastic band snapping back when it is released. But if there is enough energy in the collision, the quark will break the bonds that bind it to the other quarks, breaking free, like an over-stretched elastic band snapping in half. But does this mean that we now have a free quark? No! For by 'enough energy' we mean that there is so much energy in the collision that it can create a pair (at least) of new quarks, one on each side of the 'break' in the 'elastic band' (really, the strong force) trying to hold the recoiling quark in place. Instead of a single quark escaping, you have a pair of quarks (forming a meson), or even a new triplet; instead of two quarks from the original triplet being left behind, a new companion appears on the other side of the 'join' and remains alongside them.

This is a slightly oversimplified picture. At very high energies, instead of a simple break with one new quark appearing on each side of the join, the process of breaking the grip of the strong force would produce a shower of new particles, manufactured out of pure energy, forming a jet moving in the direction of the escaping quark. But what matters is that nowhere in that jet of particles emerging from the site of the original collision is there an isolated quark; the particles in the jet are formed out of a train of quark pairs and triplets created by repeatedly breaking the bonds between other quarks.

From 1972 onwards, experimenters at CERN were able to observe such jets in collisions between beams of particles travelling in opposite directions; this is exactly the 'colliding pancake' situation that Feynman had described theoretically a few years earlier. Throughout the 1970s, researchers at CERN and elsewhere found more and more examples of this kind of behaviour, as they probed to higher and higher energies. The important point is that the jets can come out of the collisions almost at right angles to the line of flight of the colliding pancakes, and this is only possible because at the moment of collision the quarks hardly feel the strong force restraining them at all. When they are close together, they do not notice that they are confined (a property known as asymptotic freedom); it is only when they try to escape that they feel their restraint.

Richard Field, a postdoctoral researcher at Caltech, became interested in the properties of these quark jets, and persuaded Feynman to join him in investigating the jet properties theoretically. Using the language of what is now known as quantum chromodynamics (QCD), involving quarks exchanging gluons in an analogous way to electrons exchanging photons in QED (see Figure 15), and with asymptotic freedom included as part of the package, Feynman and Field were able to make predictions about the kind of jets that should be observed. According to Field,[13] Feynman kept them honest by insisting that they only calculate the behaviour of jets in experiments that had not yet been carried out, so that the experiments

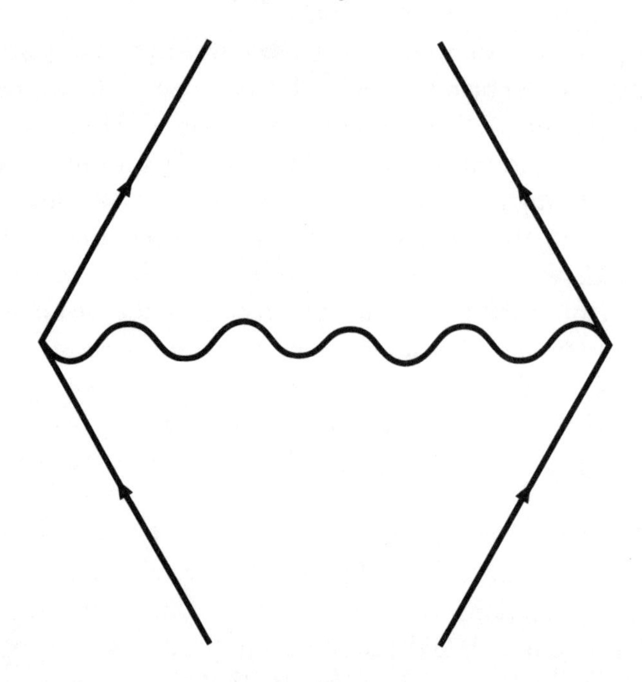

Figure 15. Also using QED as its template, QCD describes interactions between quarks. Here, two quarks on diverging paths exchange a gluon and are pulled back towards one another.

would provide a genuine test of the theory; as the higher-energy experiments were carried out, they produced jets of exactly the kind that the two Caltech theorists had predicted.

This work, some of which also involved another theorist, Geoffrey Fox, was being carried out in the second half of the 1970s. As Bjorken puts it, 'as the evidence for QCD grew, Feynman (with Richard Field) worked out the modifications to the "naive" parton model phenomenology implied by QCD, and grappled with the fundamental properties of QCD that might explain confinement'. Some of the work with Field was published in 1977, and some in 1978 – the year of Feynman's sixtieth birthday. Physicists simply don't make major contributions to their field in their late fifties, yet here was Feynman, still (or again) at the heart of new developments in particle physics. It wasn't just that his own theory of QED, developed more than 30 years before, provided the archetype on which QCD was based, but that Feynman himself was actively involved in establishing QCD as the best theory we have of the strong interaction.

But even Feynman could not go on for ever, and it was also in 1978 that he had his first encounter with cancer. There would be no more ground-breaking achievements in theoretical physics; but, as we shall see in the rest of this book, Feynman was still far from finished as an original and influential thinker. And even in the 1970s, while making his last great contribution to physics, he had, as ever, found time to follow up his fascination with science and zest for life along distinctly unconventional trails.

Notes

1. Interview with JG, April 1995.
2. Interview with JG, October 1995.
3. Yuval Ne'eman and Yoram Kirsch, *The Particle Hunters* (Cambridge University Press, 1986).
4. See discussion in John Gribbin, *In Search of the Big Bang* (Bantam, London, 1986).
5. See note 3.
6. Murray Gell-Mann, *The Quark and the Jaguar* (Little, Brown, New York, 1994).
7. This means that nobody has yet won a Nobel Prize for predicting the existence of quarks, even though Jerome Friedman, Henry Kendall and Richard Taylor shared the 1990 prize for their experimental confirmation that quarks exist! Feynman himself nominated Gell-Mann and Zweig at least once, and other physicists have done so in the 1990s; there is still time for the Nobel Committee to rectify the omission.
8. *Surely You're Joking.*
9. Bjorken, in *Most of the Good Stuff.*
10. Mehra.
11. Bjorken, personal communication to JG, October 1995.
12. R. Feynman, M. Kislinger and F. Ravndal, *Physical Review*, volume D3, page 2706, 1971.
13. Quoted by Gleick.

— 11 —
Father figure

Although he continued with his fundamental work in physics, by the 1970s Feynman was very much a family man. Even here, though, he did not always follow a conventional path. In Gweneth he had found a kindred spirit with a love for adventure to match his own, and the presence of the young children did not deter them from taking exotic and adventurous holidays. In 1973, at the suggestion of Richard's close friend Richard Davies, a physicist who worked at the Jet Propulsion Laboratory in Pasadena, they took a spring break in Mexico, visiting Copper Canyon. The plan was to take a train to a remote region of the country, then walk, with backpacks, for two or three days to a village called Cisneguito.[1]

Davies accompanied the Feynmans on this trip, acting, as he put it, as 'beast of burden'; this was just as well, because not long before they left Richard had fallen and broken a kneecap, which made walking difficult. It was after he recovered from this injury that he took up jogging to keep fit. In the village, there was a tiny schoolhouse – but the children who showed it to the visitors explained that it was no use, because they didn't have anyone to teach them. Feynman immediately took up the challenge, and began explaining to the enraptured audience, using the Spanish he had learned long before when he had first planned to visit South America, how light works. He borrowed a magnifying glass from Davies to show how the lens affects light, and held the audience in his grip as easily as he did the students at Caltech. 'I don't know', said Davies, 'if he could ever take a complete holiday from physics.'[2]

That sentiment was echoed by Michelle Feynman, recalling her childhood in the 1970s, in her contribution to *No Ordinary Genius*. 'You could never separate my father from physics', she said, commenting that 'he doodled all the time – on the edges of newspapers, on Kleenex boxes in the car . . . it seemed very strange, you know, almost a stream-of-consciousness kind of physics, pouring out of him. He had to write it down, and then he could go on to something else. So yes, every Kleenex box, every spare scrap of paper, had some sort of physics on it.'

When the historian Charles Weiner interviewed Feynman about his life and work, Dick told him much the same thing. When Weiner casually remarked that Feynman's notes on partons were 'a record of the day-to-day work', Feynman retorted, 'I actually did the work on the paper', explaining 'it's not a *record*, not really, it's *working*. You have to work on paper, and this is the paper. Okay?'[3]

Whether he could leave physics behind or not, the short trip to Mexico was such a success that they all went back for a longer visit in the autumn of 1973. This would involve a longer hike through Copper Canyon – which is deeper and longer than the Grand Canyon in the United States – visiting even more remote communities. Feynman spent some time in the summer preparing for the trip by learning a little of the language of the Raramuri, the people who lived in these remote villages, and Davies recalls how when they encountered one of the locals on the road Feynman was indeed able to communicate, after a fashion, sitting by a little fire with the Raramuri man for hours, exchanging little presents and learning each other's names. 'Richard had a kind of gift that way, of communicating whatever the circumstances. It was a great experience, and I think it illustrates the way he went about things in this straightforward, somewhat naive way.'

It was also around this time, in the early 1970s, that Feynman's longstanding interest in the culture of Central America and in codes enabled him to give another virtuoso demonstration of his communication skills, back at Caltech. Some 20 years earlier, on honeymoon with Mary Lou, he had visited a museum in a little Guatemalan town where one of the exhibits was a copy of the Dresden Codex. The Dresden Codex was a Mayan book which had been looted by the European conquerors of the New World and turned up in a museum in Dresden (at least it wasn't burned, like nearly all the other Mayan books); it is a kind of almanac and astronomical reference book, giving information about the Mayan calendar and their observations of the heavens. Because much of this information is in the form of numbers and tables, it had been possible to crack the code and translate the document.

The museum had copies of the codex for sale, with the original Mayan version on one page, and a translation into Spanish on the opposite page. It was a challenge Feynman, bored with following Mary Lou around to look at pyramids in the steamy jungle, could not resist. In *Surely You're Joking*, he tells how he bought a copy of the codex, determined to crack the code – a system of bars and dots – for himself. Covering up the Spanish translation with a piece of paper, he spent hours in their hotel room happily deciphering the code for himself, while Mary Lou climbed up and down pyramids all day (Davies was right – even on his honeymoon Dick couldn't take a complete holiday from science!).

The fun continued in Feynman's spare time back at Caltech. Eventually, he had done as much as he could. He had quickly found that a bar in the Mayan notation was equivalent to five dots, what the symbol for zero was, and how the numbers were added and carried over from one calculation to the next. He found a place in the codex in which the number 584 was very prominent, and identified this with the period of Venus as seen from Earth – 584 days, to the nearest whole number of days. Obviously, Venus was an important object to the Mayans. The 584 was divided up into intervals of 236, 90, 250 and 8 days, which could be explained in terms of the time taken for Venus to pass through its different phases, and another prominent number, 2920, could be interpreted both as 584 × 5 (five Venus 'years') and as 365 × 8 (eight Earth years), giving it double significance. A table with periods of 11,959 days turned out to be useful for predicting eclipses, but there were other numerical relationships that Feynman only figured out much later, and some which nobody has yet figured out at all.

So Feynman at last turned to the Spanish translation to see how his interpretation compared with that of the experts – only to find that the Spanish text wasn't a translation at all, but a commentary describing some of the symbols used in the Mayan text. Feynman had to follow up his continuing interest in the Mayans elsewhere, especially in the books of Eric Thompson, and his interest became known to a few of the experts in the field.

In the 1970s, this interest was rekindled. One of the professors at the University of California, Los Angeles (UCLA), Nina Byers, had just taken over organizing the weekly meetings known as colloquia, where physicists from other universities usually come to talk about their work. She decided that it would be a good thing to broaden the minds of her colleagues by introducing them to subjects outside their own culture, and since Los Angeles is near Mexico, she felt that a good place to start would be to have a colloquium on Mayan mathematics and astronomy. She called a specialist, Otto Neugebauer, of Brown University, to ask if he could recommend somebody on the West Coast who could do the job – and was told that the best person in the LA area was not a professional anthropologist or historian, but an amateur, someone she might have heard of, a certain Richard Feynman.

'She nearly died!', Feynman recounts. 'She's trying to bring some culture to the physicists, and the only way to do it is to get a physicist!'[4]

By then, Feynman had lost his copy of the Dresden Codex, and when Byers bit the bullet and asked him to give the talk she provided him with a new, clearer copy, so he could reconstruct his calculations. This time he went a little further than he had in the 1950s, discovering that some of the strange numbers he hadn't understood earlier were an attempt by the Mayans to get closer to the true Venus cycle of 583.92 days instead of the round 584 days.

The colloquium was such a success that Feynman was asked to give the same talk a little later at Caltech. Shortly before the date of the Caltech lecture, news broke of the discovery of a new codex (only three of these Mayan documents had ever been discovered), and Feynman got hold of a picture of the fragment of supposedly Mayan writing to describe in his talk. He quickly spotted that it was a fake – it used the same numbers as in the Dresden Codex. The odds against two out of four surviving fragments from the vast Mayan literature both referring to the orbit of Venus are so great that the new codex had to be a fake. It was as if the entire Library of Congress had been burned to the ground, with fragments of only four books surviving, and two of them turned out not only to be pages from different editions of the same almanac, but pages from the same chapter of each book!

Feynman was disappointed at the lack of courage and imagination of the hoaxers. He could have done it so much better:

> A real hoax would be to take something like the period of Mars, invent a mythology to go with it, and then draw pictures associated with this mythology with numbers appropriate to Mars – not in an obvious fashion; rather, have tables of multiples of the period with some mysterious 'errors,' and so on. The numbers should have to be worked out a little bit. Then people would say, 'Geez! This has to do with Mars!' In addition, there should be a number of things in it that are not understandable, and are not exactly like what has been seen before. That would make a *good* fake.[5]

Feynman got a big kick out of giving this talk. 'There I was, being something I'm not, again.' The other big thing that he got a kick out of through being 'something I'm not' (apart from drawing) was his drumming. He had originally been self-taught, playing by instinct and copying rhythms he heard on records of African drummers. At Cornell, he had had lessons, and he learned bongo rhythms in Brazil. After he moved to Caltech he had met a Nigerian drummer called Ukonu, who played in a nightclub on LA's trendy Sunset Strip. Ukonu was a medical student, but a sufficiently talented drummer to have made professional recordings; he gave Feynman some rather chaotic tuition in his own style, and opened up the opportunity to jam with other drummers. Ukonu went back to Nigeria a little before the beginning of the civil war there in 1967, and Feynman never heard from him again. After that, the drumming lapsed a little, while Feynman concentrated on other things (this was about the time he was working on partons, when Michelle was a baby). But in the 1970s it flowered in a quite unexpected way, thanks to his friendship with the Leighton family.

Robert Leighton was a long-time colleague of Feynman, and had worked with him on the *Lectures*. At a dinner party at the Leightons' house, Feynman discovered that Robert's son, Ralph, and Ralph's friend Tom Rutishauser were keen drummers, as well as being what Dick called 'real musicians' – Ralph played piano, and Tom the cello. Ralph recalls that he was around 17 at the time, and that although Feynman had been 'introduced' to Ralph as a baby (and had given him an old typewriter when Ralph was six), this was the first time Ralph really became aware of Richard.

> We were at a very impressionable age, in high school, getting tired of our parents telling us what to do, but unconsciously looking for some kind of role model who had been around. So here's this guy who liked to drum, who had these incredible stories – he cracked the safe that had the secrets of the atomic bomb! This was at the time of the Vietnam War, and he had a story about the draft (a subject of more than passing interest to us): this atomic scientist was rejected by the army as mentally *deficient*! Tom and I were totally fascinated. I realise now that in the environment I grew up in, the 'cultural wasteland' of middle-class America, there was no storytelling tradition. And I also see now that his mother's way of telling stories – and her sense of humour, her appreciation of irony and absurdity – were all part of a crucial side of him, an essential element of Feynman.[6]

The growth of that storytelling into two books came much later. At first, the three of them started drumming together once a week, and worked out some good rhythmic patterns. They progressed to playing at schools, providing the rhythms for a dance class and doing other odd gigs under the name 'The Three Quarks'. Then, after Tom moved to the East Coast to pursue his career as a cellist, Richard's drumming took a new turn.

It started when Feynman was asked to play a small part, as a bongo player, in a Caltech production of *Guys and Dolls*. As is often the case with such campus-based amateur theatrical companies, there was a tradition of roping in eminent members of the faculty to play bit parts in the productions, and since there is a nightclub scene in *Guys and Dolls*, on this occasion the director thought it would be fun to have Feynman playing the part of a musician in the nightclub. Feynman agreed readily, but was petrified to discover that he was actually supposed to read music and play a prearranged drum piece to fit in with the storyline. Since he didn't read music, the problem looked insurmountable, until he brought Ralph in to interpret the notation for him and teach him his part. Soon, Ralph was enrolled to play the part of another musician in the nightclub scene, and together they pulled it off, to the delight of the audience.

The same scene in the nightclub also involved some dancing, and the wife of one of the faculty at Caltech happened to be a choreographer working for Universal Studios, so she had been roped in to organize the dance. She liked the combined drumming efforts of Ralph and Dick, and to their astonishment asked them to drum in San Francisco for a ballet she was going to choreograph there. The good thing was she didn't want them to play prearranged music, but intended to listen to their drumming, tape the segments she liked, and use that as the basis for the choreography. The drumming for the show, though, would be live, not prerecorded; and there would be no other musicians involved.

Ever eager for new adventures, Feynman had no trouble persuading Ralph to go along with the idea, but insisted that nobody involved in the San Francisco project should be told that he was a famous professor of physics. If he was going to drum professionally, he wanted to be taken strictly on his merits as a drummer. It always baffled Feynman when he was introduced to an audience as a professor of physics who also played the drums – this happened, for example, when he gave the Messenger Lectures, which were turned into the book *The Character of Physical Law*. Meaning well, but to Feynman's irritation, the Provost of Cornell University introduced Feynman on that occasion by commenting that 'my Caltech friends tell me he sometimes drops in on the Los Angeles night spots and takes over the work of the drummer' (this was in 1964, during Feynman's friendship with Ukonu). That is why the first of those Messenger Lectures begins with an unrehearsed comment:

> It is odd, but on the infrequent occasions when I have been called upon in a formal place to play the bongo drums, the introducer never seems to find it necessary to mention that I also do theoretical physics. I believe that is probably because we respect the arts more than the sciences.[7]

When the time to drum for the ballet came around, in November 1976, it all worked out – but with some unexpected difficulties, as Feynman describes in *Surely You're Joking*. Nobody involved realized he was anything other than a professional drummer, and although the audience was small (about 30 people all told) both they and the dancers were appreciative of the drumming. And he did indeed get paid for the work. 'For me, who had never had any "culture", to end up as a professional musician for a ballet was the height of achievement, as it were.'

All this time, in and around his research, his trips abroad and his drumming, Feynman was more than ever a kind of icon or guru to the undergraduates at Caltech. David Goodstein recalls[8] that for the best part of two decades Feynman gave an informal 'course', known as Physics X, to a class which met every week at 5 o'clock on a Monday or Tuesday afternoon. There were no credits for attending, and no set curriculum, but the room was always full. Feynman simply discussed whatever the students wanted him to discuss, and the only rule was that no members of the faculty were allowed to attend. Many students felt that it was like having a hot line to God, as Feynman always attempted to explain even the most esoteric ideas in physics in a clear-cut, down-to-earth manner. Alas, because of its informal nature, no record of exactly what went on in Physics X was ever kept.

The students also had unlimited access to Feynman, whenever he was around, on a one-to-one basis. Just as in the old days at Cornell, as Dyson had learned back in the 1940s, if Feynman was really busy on some tricky aspect of physics the casual visitor to his office would be greeted with a shout of 'Go away, I'm busy.' Otherwise, though, his secretary Helen Tuck (who worked with him from 1971 onwards) had unconditional instructions that he was always available to any student who wanted to see him.

He sometimes had less time for more senior members of Caltech. Helen Tuck's office was between the offices of Murray Gell-Mann and Richard Feynman, and she looked after both of them. The door to her office is on the right-hand side of the wall, facing her desk, and it just happens that there is a structural pillar to the left of the door, inside her room, so that the chair used by visitors, its back against the wall, is not in sight from the doorway. Often, Feynman would sit in that chair, chatting to her about life in general, when he wasn't in the mood to work. Sometimes, a visitor would come to the door, asking, 'Is Professor Feynman in his office?' She would glance at Dick, and if he shook his head she would reply, truthfully, 'No, he isn't in his office just now', and the visitor would go away. It was a harmless way to avoid being imposed on when he wasn't in the mood, calculated to avoid giving offence (much worse to be told, 'Yes, he is in but he doesn't want to see you'), and Tuck and her colleagues were deeply hurt when one biographer described how Feynman 'hid behind her door' to avoid being seen; the comment, she felt, displayed a fundamental lack of understanding of Feynman's character.[9]

Even with the best intentions, though, Feynman found it very difficult to be responsible for research students. He said, 'I've put a lot of energy into my students, but I think I wreck them somehow. I have never had a student that I felt I did something for, and I have never had a student who hasn't disappointed me in some way. I don't think I did very well.'[10] As this comment shows, Feynman blamed himself, not the students, for what he perceived as their failure to make a mark in science. Part of the problem, as we have mentioned, was that Feynman couldn't resist solving problems. If he found a good problem for a student to work on, he'd end up solving it himself; if students came to him with a problem, he couldn't help but solve it for them, rather than just giving them enough of a hint to get them going in the right direction to solve it themselves. He didn't mean to, but he couldn't help it; present Feynman with a problem, from a Mayan codex to a locked safe to the mysteries of quantum electrodynamics, and he would just *have* to solve it – the exception that proves the rule, of course, being his promise to his sister Joan to leave the aurora to her. Hans Bethe, Feynman's old mentor, had the same problem. So while some great physicists, such as Oppenheimer, produced a stream of doctoral candidates who had learned to do physics the Oppenheimer (or whatever) way, and carried the style of their teacher forward into the next generation, there was never a 'school' of Feynman students in the same sense.

Another problem was that Feynman made no concessions to students. He treated everybody the same way. When he was a youngster at Los Alamos, he did not hesitate to tell Bethe he was a fool if the older man made a mistake; now that he was a senior scientist, he did not hesitate to tell his students (or anyone else) they were fools if they made a mistake. It was no more, and no less, than he expected himself – indeed, he would often describe his own errors as foolish, or stupid, mistakes. But it is hard for graduate students to cope with that kind of criticism from their supervisor. One Caltech student who went on to achieve eminence in the field of relativity theory, Kip Thorne, says that as a young researcher he was terrified of giving a seminar when Feynman was in the audience.[11] But although it could be distressing to have Feynman bluntly pointing out the flaws in your argument, as another former Caltech student pointed out to us[12] it was, ultimately, always acceptable, for one important reason. Feynman was always right. He could see the faults in an argument quicker than other people could. If he said there was a flaw in the argument, there was; and it was surely better, when it came down to it, to find out from him, before you made a complete fool of yourself by publishing the mistake in a journal for all the world to see. It was also a good idea to be careful what you wore to give a seminar at Caltech, especially if there were flaws in your argument. Feynman's dislike of uniforms and authority could encourage him to even stronger attacks on anyone who seemed to be trying to pull rank; 'if somebody came to give a lecture in a suit, he would be merciless'.[13]

The students who did do well (in spite of Feynman's disclaimer, there were some) were the ones who quickly learned that the abrupt dismissal of bad ideas was not meant personally, and who worked through the curtness without taking offence. You needed initiative in order to convince Feynman that you were worth spending time on. There's another important point. According to some of those who worked alongside him, it wasn't so much that Feynman was a failure with graduate students as that he had relatively few of them. This was because he didn't work at the head of a large group, but mainly on his own, so that when an interesting problem occurred to him it was natural to get on with it himself, rather than pass it on to other members of a team.[14]

One of the best examples of how to succeed as a research student with Feynman comes from Michael Cohen, who explained his approach in *Most of the Good Stuff.* Cohen graduated from Cornell in 1951, but had scarcely known Feynman there. He moved to Caltech to work for his PhD, hoping to work with Feynman, who was away in Brazil during Cohen's first year on the West Coast. They got to know each other when Feynman returned to Pasadena, and Cohen made a deliberate effort to make himself useful by studying Feynman's own papers on liquid helium, and looking for areas in which the work could be extended, rather than simply going to Feynman and asking for a problem to work on. This led to a genuine collaboration, and Cohen also learned much about intellectual honesty from Feynman.

As Cohen's thesis adviser, Feynman worked through all Cohen's calculations in the first draft of the thesis, and found a numerical error. With the error in place, the calculation gave almost perfect agreement with a number determined by Lev Landau; with the correction, Cohen's result was 20 per cent higher than Landau's. Just because the first calculation had seemed to give the 'right' answer didn't mean that it shouldn't be checked, and the honest result was the one that appeared in the final version of the thesis. Cohen stayed on with Feynman for 18 months after completing his PhD work, until 1957. Then, at Feynman's recommendation, Oppenheimer took him on at the Institute for Advanced Study.

Feynman cannot have been *too* disappointed with this particular student if he recommended him to Oppie! So no matter how gloomy Feynman may have felt about his track record with research students on the day in 1988 that he discussed them with Mehra (only shortly before his death), there were some successes, and his comments on that occasion should be taken with a pinch of salt, as rather extreme self-criticism.

The problem with most research students was that they fell into a gap between two kinds of scientist who could benefit from Feynman's unique ability. Undergraduates could benefit from contact with him precisely because he was a kind of oracle who could fill their heads with ideas and images about the wonderful world of physics. Graduate students had trouble because he couldn't give them space to develop their own ideas. But his peers in research, who had already found their own space, could benefit from precisely the trait that caused trouble with many research students, his compulsion to solve puzzles. If anybody was stuck with how to develop an idea in physics, they only had to call Dick Feynman and he would point the way through the immediate logjam so that they could get on with their work.

As Willy Fowler told Mehra, 'you just had to tell him a few lines, and he would jump up with ideas and diagrams. He was very helpful and encouraging. Feynman was interested in everything . . . He was just tremendous.'

Another physicist, Richard Sherman, saw just how tremendous Feynman could be in this problem solving role, when Sherman was halfway through his first year as a graduate student at Caltech, doing research on superconductivity. He was in Feynman's office, writing up equations on the blackboard, and Feynman was analysing the work almost as quickly as Sherman could write. Then, the telephone rang. The caller had a question about a problem in high-energy physics. Feynman immediately switched into a discussion of the complicated problem involved, talked for about 10 minutes and resolved the caller's difficulty. He hung up the phone, switched back to superconductivity and carried on exactly where he had left off, until the phone rang again. Somebody else had a problem, involving solid state physics. Feynman solved it, and went back to superconductivity again. 'This sort of thing went on for about three hours – different sorts of technical telephone calls, each time in a completely different field, and involving different types of calculation. [It] made a tremendous impression on me. It was staggering. I have never seen that kind of thing again.'[15]

Another Caltech graduate student, who was supervised by Murray Gell-Mann in the 1960s, unconsciously echoed Marc Kac's comments about the nature of genius (which he was unaware of at the time) when he told us that 'Murray was clever, but you always had the feeling that if you weren't so lazy and worked really hard, you could be just as clever as him. Nobody ever felt that way about Dick.'[16] Feynman may not have built up a large school of graduate students under his direct supervision, but he was a father figure and inspiration to all the graduate students in physics at Caltech during his time there, even the ones supervised by Gell-Mann!

Hagen Kleinert, who now works at the Institute for Theoretical Physics in Berlin, visited Caltech as a young professor in 1972. 'I had actually been hired by Gell-Mann,' he told us, 'but he was very hard to learn from since he always pretended to know everything from pure intuition without any ditch work.'[17] The person Kleinert learned most from during his visit was Feynman, who gave a weekly seminar on the path integral approach to the young postdoctoral researchers. During the course of these seminars, Feynman explained that he had stopped teaching path integrals at a less advanced level, because he had never derived a complete path integral description of the hydrogen atom, and was embarrassed by this failure. The path integral idea provided a superb mental picture to give a physical feel for what is going on, but the calculations had proved intractable. Actually, this was no real disgrace. The standard approach to quantum mechanics, using Schrödinger's wave equation, was not much better, since even the Schrödinger equation could only be solved to give an exact description of hydrogen, the simplest atom of them all.

The idea stuck in Kleinert's head, and several years later he not only solved the problem (much to Feynman's delight), but wrote a major textbook on the path integral approach, re-establishing path integrals as a research tool, not only conceptually useful but now capable of solving problems as easily as using the Schrödinger equation.

In 1982, Kleinert was back in California (this time based at Santa Barbara), and visited Caltech several times. 'Feynman knew of my work on the path integral of the hydrogen atom by then, and was very friendly to me and open to discussion.' The friendship extended to some joint work, updating some of Feynman's earlier ideas with the aid of a Sinclair home computer, one of the first computers available to the public, that Kleinert had just bought at Woolworth's for $15.00. At first, the work seemed of only minor importance. But in the 1990s Kleinert and his colleagues have developed the technique, known as the variational principle, into a powerful tool which makes it possible to use path integrals to solve increasingly difficult problems in the quantum world. And it all stems from Feynman's continuing active involvement in fundamental science, as a father figure pointing the way for younger researchers, well into the 1980s.

Feynman was also a father figure to the undergraduates. In the 1974 commencement address, which we mentioned in Chapter 10, he provided them with words of wisdom about science which were also words of wisdom about life in general, just the sort of thing a father ought to pass on to his children before they go out into the world. Shooting down the widespread public acceptance of what he regarded as pseudosciences like astrology and spoonbending[18] (and, one of his eternal bugbears, psychology), he explained what it was that real science had that these pseudosciences did not:

> It's a kind of scientific integrity, a principle of scientific thought that corresponds to a kind of utter honesty – a kind of leaning over backwards. For example, if you're doing an experiment, you should report everything that you think might make it invalid – not only what you think is right about it: other causes that could possibly explain your results; and things you thought of that you've eliminated by some other experiment, and how they worked – to make sure the other fellow can tell they have been eliminated.
>
> Details that could throw doubt on your interpretation must be given, if you know them. You must do the best you can – if you know anything at all wrong, or possibly wrong – to explain it. If you make a theory, for example, and advertise it, or put it out, then you must also put down all the facts that disagree with it, as well as those that agree with it.[19]

Few scientists have such complete integrity. Even the most honest subconsciously cut the odd corner, or neglect to mention all of the evidence in conflict with their pet theory. But Feynman never succumbed to his own wishful thinking. He never fooled himself. In Fowler's words, 'Feynman was a very wise man, who set very high standards for everyone. He motivated you to achieve them. Just the fact that he was around, all of us at Caltech thought that we had to live up to his standards. In this indirect way he influenced us all.'[20]

The fatherly, 'wise man' influence extended outside the campus, to Feynman's wider circle of friends and acquaintances. By the end of the 1970s, he had yet another outside interest to add to his list, one that was to be a recurring preoccupation in the last decade of his life. In the summer of 1977, he had just about finished his work on quark jets, Carl was soon to be beginning his junior year at the local high school, and Michelle had completed first grade.[21] Ralph Leighton had a job teaching mathematics at the same Pasadena high school where Carl was a student, but confessed to Feynman over dinner one day that what he would really like would be to teach geography. Feynman responded by asking Leighton if he had ever heard of a place called Tannu Tuva, which Feynman knew from his childhood hobby of stamp collecting. No philatelist himself, and convinced that Feynman was pulling his leg, Leighton insisted that they look for it in the atlas at the back of the *Encyclopaedia Britannica.* What they found, nestling to the northwest of

Mongolia, was a tiny region labelled 'Tuvinskaya ASSR', part of the Union of Soviet Socialist Republics. Ralph conceded that the region could once have been called Tannu Tuva, after noting the Tannu Ola Mountains to the south. When they discovered that the capital of the country was called Kyzyl, a name completely without a vowel in it, there was only one reaction:

> 'We must go there,' said Gweneth.
>
> 'Yeah!' exclaimed Richard. 'A place that's spelled K-Y-Z-Y-L has just *got* to be interesting!'
>
> Richard and I grinned at each other and shook hands.[22]

At the time, the problems of visiting such a remote region of the USSR seemed insurmountable, which made the project all the more appealing. Of course, Feynman would have been able to arrange an official lecture tour of the Soviet Union, with a guaranteed trip to Kyzyl as part of his payment. But that would have been too easy. Just as he wanted to be treated on merit as a drummer, not regarded as some kind of freak spectacle, a 'physicist who drums' (like a dog walking on its hind legs), so he wanted to visit Tuva as an explorer, if only he could find the way through the red tape and bureaucracy that would inevitably confront an American citizen (who happened to have worked on the atomic bomb) trying to visit a communist country (and one far off the tourist trails) during what was still then the Cold War.

They made slow progress with the project, mostly because for a long time it wasn't a serious endeavour, and partly because it was around this time that Feynman first became seriously ill with cancer.

Richard and Gweneth were on holiday in the Swiss Alps – one of their regular haunts – in the summer of 1977 when Richard first showed signs that something was wrong. There may have been symptoms before, which he had ignored, having other things on his mind; but on this occasion he frightened Gweneth by suddenly running into the bathroom and vomiting.[23] Although clearly ill, Richard pushed his own health worries to one side, not least because of his concern about Gweneth, who had cancer and had to undergo an operation. So it was only at the end of the summer of 1978, when he went to the doctor complaining of abdominal pains, that his own cancer was diagnosed. By then, it was 'a fourteen-pound mass of cancer the size of a football',[24] and showed as a visible lump at Feynman's waist. Growing in his abdomen to this enormous size, the tumour had crushed Feynman's left kidney and adrenal gland, and had wrecked his spleen.

One day, Helen Tuck phoned round to Feynman's colleagues at Caltech, including David Goodstein, to inform them that Dick had cancer, and would be undergoing major surgery the following Friday. On the Monday before the operation, Goodstein recalls,[25] he mentioned to Feynman that there seemed to be an error in a piece of work they had done together. It was nothing of great importance, but the work had been published, and they ought to set the record straight. Feynman agreed to look into it, and was soon absorbed. 'He didn't know whether he was going to live through the week, but here he was absorbed by a really not important problem in elastic theory.'

At the end of the afternoon, they decided that the problem couldn't be solved, and went home. Two hours later, Goodstein had a phone call from Feynman. He was on a complete high, absolutely exhilarated, because he had found the solution to the problem, which he dictated to Goodstein there and then. It was four days before the operation, and the problem was still unimportant, but solving it had made Feynman's day. 'I think', says Goodstein, 'that tells you a little bit about what drove the man to do what he did.'

The operation, performed at the local hospital in Pasadena, seemed to be a success, but after it Feynman, although still cheerful, was not only physically affected by the damage done to him by the cancer, but also knew that he was now living on borrowed time. Of course, all of us are under sentence of death, in the long term. But in Feynman's case, the future had begun to close in. Subscribing to his own philosophy of 'never fool yourself', he treated the cancer as an interesting case study, and looked up all he could about it, observing the changes going on in his own body like a scientist watching an experiment. The cancer, a so-called liposarcoma, was malignant, and although the huge lump had been removed, the textbooks said that there was essentially zero chance of his surviving for another 10 years.

Meanwhile, life continued as normal, including his physics, his teaching, drawing, holidays, intermittent efforts to get the Tuva project under way and drumming with Ralph Leighton. The drumming sessions, though, became as much storytelling sessions as anything else, as the tape of the 'Safecracker Suite' highlights.[26] Feynman had always been an inveterate storyteller, but now he seems to some extent to have taken stock of his life by pouring out the anecdotes that became the two books on which Leighton collaborated with him. Leighton feels that he was simply the right person in the right place when Feynman was ready to make his stories available to a wider audience. He recalls an occasion at dinner when Feynman mentioned that he had been being interviewed about his scientific work, but that whenever he got on to 'the good stories' the interviewer would shut his tape recorder off. 'Feynman was kind of complaining about that,' says Leighton, 'so I piped up that those were my favourite stories, so let's see if we couldn't write them down in an organized way.'

All things considered, it's hardly surprising that Feynman produced little in the way of new ideas in physics after his sixtieth birthday. But fate was to give him one last great opportunity to demonstrate to the world the way that a top scientist thinks, and how the scientific method should be applied to solving problems. This opportunity, reluctantly taken, also made Richard Feynman even more famous than he had been already; above all, though, it was to highlight a damning example of what can happen to organizations, as well as to people, when they start fooling themselves by believing what they want to be the truth, rather than what really is the truth.

Notes

1. Davies has told the story of this trip in his contribution to *No Ordinary Genius*.
2. See note 1.
3. Weiner interview with Feynman, quoted by Gleick.
4. *Surely You're Joking.*
5. See note 3.
6. Leighton, interview with JG, April 1995.
7. Feynman, *The Character of Physical Law*.
8. Interview with JG, April 1995; see also *Most of the Good Stuff*.
9. Helen Tuck, interview with JG, April 1995; the biography she is referring to is the one by Gleick.

10. Mehra.
11. Conversation with JG, April 1995.
12. Norman Dombey, interview by JG, October 1995. Dombey also says that the only occasion on which he was present at a Caltech seminar and Feynman didn't tear the speaker's arguments to shreds was when Oppenheimer was the speaker. But he couldn't say if that was down to Oppenheimer's brilliance, or Feynman's respect for his boss from the Los Alamos days.
13. See note 12.
14. This point was made in an interview with JG, in October 1995, by Norman Dombey, who completed his PhD at Caltech under Murray Gell-Mann in 1961, and worked there in 1961–62 as a postdoctoral researcher.
15. See Sherman's contribution to *No Ordinary Genius.*
16. The former student, now an eminent physicist in his own right, asked to remain anonymous in order not to suffer any backlash from Gell-Mann about this remark. The same physicist told us that 'Murray always had to be right, even when he wasn't, but Feynman wasn't afraid to admit when he'd made a mistake.'
17. Correspondence with JG, January 1996.
18. Interviewed by JG in April 1995, Helen Tuck recalled that Richard and Carl Feynman once had a private meeting with Uri Geller, and found that none of Geller's claimed powers operated under their careful scrutiny.
19. *Surely You're Joking.*
20. Quoted by Mehra.
21. Carl's teacher called Gweneth one day, expressing concern that Carl, who seemed a bright child, had scored only 129 in an IQ test; Gweneth told the teacher that this wasn't too bad, since Richard's IQ was only 123.
22. Leighton, *Tuva or Bust!*
23. Gweneth Feynman, as told to Gleick.
24. Leighton, *Tuva or Bust!* There is some confusion about the mass of the tumour, which Gleick says weighed 6 pounds; but there is no doubt about the damage it caused to Feynman's internal organs. Freeman Dyson, usually a reliable witness, also quotes a figure of 6 pounds, in his book *From Eros to Gaia.*
25. David Goodstein, interview with JG, April 1995; see also *No Ordinary Genius.*
26. See Bibliography.

— 12 —
The last challenge

The event that would make Feynman known to a wider audience than ever before was the explosion of the space shuttle *Challenger* in 1986. But although his work for the *Challenger* inquiry became the best known of his activities in the final decade of his life, it was far from being his only piece of technical work after he turned 60. Although Feynman made no major contributions to theoretical physics in the 1980s, he did have an absorbing scientific interest – one that harked back to his childhood fascination with solving mathematical problems, and to his work at Los Alamos in charge of the Theoretical Computations Group. With his son Carl (whose interest had, happily for Feynman, switched from philosophy to computing), he became involved in the development of the next 'big idea' in computers, parallel processing.

Carl studied at MIT, where his father introduced him to Marvin Minsky, one of the pioneers of research into the possibility of creating artificial intelligence. Through Minsky, Carl met Danny Hillis, a graduate student who had a crazy ambition, to build a giant computer. 'Well,' said Carl, 'what did I know? I was seventeen years old, and I thought it would work – nobody else did.'[1] So Carl became one of the undergraduates helping Hillis out with his thesis project.

The idea behind the plan to build a giant computer was that instead of having one huge machine (one 'central processor', in computer jargon) working on a single huge problem, you would break the problem down into smaller bits and feed each of the pieces to a smaller processor, with all the small computers linked together so that they could cooperate in taking the various calculations through to their logical conclusion. This is parallel processing, which has begun to become an important practical possibility in the 1990s. It is, of course, exactly what Feynman did at Los Alamos in the 1940s, only then his computers (the parallel processors) were human beings operating calculating machines, every person solving their own tiny bit of the problems involved in making the first atomic bombs. The dream Hillis had was of a million computers working together in this way – a million processors operating in parallel. As his dream began to look like becoming a reality in the early 1980s, he had to lower his sights a little and settle for 64,000 processors working together, 16 of them on each single computer chip, with 4000 computer chips wired together and programmed in the right way to do the problem solving. Anybody who knew Feynman could have guessed that once he heard about the project he would have to get involved.

It's no coincidence that Feynman was acquainted with Marvin Minsky. He had maintained his interest in computation, off and on, since his work at Los Alamos, and by the end of the 1970s that interest extended to the theoretical limitations of computers, as well as to the practical aspects of building them and making them work. As a result of a question posed by the head of the computer department at Caltech, Feynman had tried to discover the minimum amount of energy required, in theory, to carry out a computation, and was intrigued to discover that there is no lower limit. No matter how little energy was available, an ideal computer would still be able to carry out its work.

At a meeting on computation at MIT, Feynman was pleased to discover that a real computer expert, Charles Bennett, had reached the same conclusion. This led to discussions about the limits set by the rules of quantum physics, a puzzle worried over by several physicists, and involving a visit by Bennett to Caltech. Once again, the surprising conclusion was that there are no limits, except for the physical ones, like size. The smallest, fastest computer possible would store numbers on individual atoms, as a string of binary digits (zeros and ones) indicated by some property such as the spin of the atom (up or down), and carry out computations using those strings of numbers.

Feynman was also intrigued by the way in which the workings of artificial computers differ from the workings of the human mind:

> I found it amusing that the things I consider myself smart at – for instance, when I was young I was good at calculus, playing chess, and other logical things – could be done by computers . . . Mathematical and logical thinking, which we were always so proud of, that they can do. It's illogical thinking that . . . we do immediately, easily, as the eye jumps from one part of the scene to another and integrates the whole picture into a room with chairs and furniture and everything that we see, that's difficult [for computers]. It's very interesting. Altogether, computers are fascinating and the problems that they can do are fascinating.[2]

It's actually slightly more subtle, and fascinating, than even this example indicates. Even the things that computers do well, like playing chess, they do not necessarily do in the same way that people do them. A good computer chess program works by considering a large number of possible moves (perhaps every move it can make), looking ahead to every possible response to each move, then at each possible next move, and so on (down to a 'depth' decided by the power of the computer and the amount of memory that it has available) to decide which is the best move to play. A good human chess player looks at the whole pattern of the pieces on the board, developing a feel for the balance of power, and often deciding on a particular plan of campaign (or, just as important, rejecting an alternative plan) because it fits (or does not fit) the overall 'feel' of the game.

In spite of what Feynman said about the things he used to be good at himself, what made Feynman a great scientist was not his ability to think logically and carefully like a machine. His great achievements – for example, QED itself – came about as much through intuition as anything else, through having a 'feel' for physics, knowing instinctively (which means, as a result of this subconscious process that he talks about) what is the right approach. He never did develop a completely logical version of the path integral approach to QED and Feynman diagrams; to this day, the great successes of this approach are built upon making inspired guesses to develop a description of what goes on in some interaction, and then tinkering with the resulting diagrams and equations to make the guesses agree more and more with the real world of experiment. Feynman seemed to understand how nature must respond in different circumstances, in the same way that nature herself understands. A ball following a curved trajectory through a window doesn't have to calculate a complicated mathematical equation in order to follow the path required by the Principle of Least Action, and Feynman didn't have to invent a rigorous mathematical proof in order to know that his version of QED worked. He was, indeed, a magician, not an ordinary genius.

Feynman was also attracted by crazy ideas. If everybody worked in the same safe areas of conventional research, after all, progress would be very slow. He always encouraged people to try out wacky ideas, because although the chance of any one of the ideas being fruitful might be small, the potential rewards for anyone who did hit the jackpot would be enormous (of course, you had to know where to draw the line, and Feynman did not encourage people to pursue wacky ideas that disagreed with experiment; this was not an endorsement of spoonbending or ESP). So when, in the spring of 1983, Hillis told Feynman that he was planning to leave the MIT Artificial Intelligence Lab and start a company to build a computer using a million parallel processors, the reaction he got – 'That is positively the dopiest idea I ever heard'[3] – was actually a ringing endorsement of the plan. Over lunch, Feynman agreed (perhaps 'insisted' would be a better description) that he would spend his summers working for the company, as yet unnamed, that Hillis planned to set up. Apart from the fun of new problems to solve, it would give him more time with Carl.

Although delighted to have a Nobel laureate on his letterhead (when he got around to having a letterhead), Hillis had no real idea what to do with Feynman. When Dick arrived in Boston that summer to start work, the company had only just been incorporated, and was largely staffed by young people who had not yet formally graduated from MIT, although they had finished their courses there. When he asked them what his job was, after some discussion they told him that he could advise them on the application of parallel processing to scientific problems. He was having none of that. 'Give me something real to do', he said.[4] So they sent him out to buy some office supplies, and when he got back they told him that he could analyse the way in which the individual processors would communicate with each other – a system known as a router, which would be responsible for finding a way for each communication between individual processors to travel along the wires linking them into one machine, without interfering with other messages travelling along the wires.

Feynman focused intently on the problem, but also found time to help out in wiring up the machine, setting up the machine shop and shaking hands with investors in the project. He also made a major contribution to setting up the structure of the company, encouraging Hillis to set up different teams, each under a group leader, working on specific tasks, just the way things had been done at Los Alamos (itself, in effect, a form of parallel processing). Just about every facet of his lifetime of experience turned out to be relevant to something that was going on in the project.

By the time he had completed his main task, analysing the requirements of the router, the company had a name – Thinking Machines Corporation – and so did the machine – the Connection Machine. Feynman's analysis showed that in order to work efficiently, each of the chips in the Connection Machine would require a minimum of five buffers for its communications with the rest of the machine, to prevent a logjam of messages piling up. Conventional computer wisdom had it that they would require seven buffers per chip, and in order to play safe the team decided to go with the conventional wisdom. But when it became time to make the chips, it turned out that they were too big to be manufactured using standard technology. With five buffers on each chip instead of seven, the manufacturing would be straightforward. Hoping that Feynman was right, they went ahead with the smaller design. It worked, and the first program was successfully run on the Connection Machine in April 1985.

By then, Feynman had made many more contributions to the project. He showed the young team the importance of cutting out jargon and explaining their work clearly, using everyday language wherever possible, when describing it to other people (including those investors). He soldered circuit boards, and helped paint the walls. Meanwhile, at Caltech, a conventional computer was being built to carry out computations simulating what happens when quarks interact with one another, and Feynman wondered whether the Connection Machine (which had not yet been completed) could do that. He made up a computer program that could tackle the job using the principles of parallel processing, and then worked through some of the steps in the calculations that would be involved on paper, to see how much processing power would be needed to do the real job, and how long it would take. He was, in fact, simulating with his paper and pencil the operation of a computer simulating the interactions between quarks using the rules of quantum chromodynamics. He found that it would work – the Connection Machine, when completed, would be able to carry out calculations involving QCD faster than the conventional machine being built at Caltech specifically to carry out calculations in QCD! 'Hey, Danny!', he yelled. 'You're not gonna believe this, but that machine of yours can actually do something *useful*!'[5]

In *Most of the Good Stuff*, Hillis describes the last piece of work he did with Feynman, a simulation of the way in which populations of living creatures evolve, in accordance with the Darwinian principle of Natural Selection. Hillis had been surprised to discover that in computer simulations populations seemed to stay fairly stable for many generations, and then to evolve suddenly into new forms. This echoes the appearance of many features of the fossil record, which has led to a variation on the Darwinian theme known as punctuated equilibrium. Together with Feynman, Hillis worked out a theory to explain this, a mathematical model of evolution at work. Then he discovered that it had all been done before and that biologists already knew about it. Disappointed, he called Feynman to pass on the bad news. But Feynman was elated. 'Hey, we got it right! Not bad for amateurs!' As ever, what mattered to him was the pleasure of solving the problem *himself*. He didn't care whether someone else had solved it first.

Feynman was the ideal person to work on the Connection Machine, a father figure to the team, because, as Hillis says, 'he was always searching for patterns, for connections, for a new way of looking at something'. But 'the act of discovery was not complete for him until he had taught it to someone else'.

By the middle of the 1980s, Feynman was seriously ill again, and his friends knew that he could not have long to live. But he was to have one last opportunity to find a new way of looking at something, to make connections, and, best of all, to explain his discovery to a large audience. Sadly, though, the opportunity came about as a result of a human tragedy which stunned the entire country.

The *Challenger* disaster occurred shortly before noon, Eastern Standard Time, on Tuesday, 28 January 1986, when the space shuttle exploded, a little over a minute after lift off, killing all seven crew members. Feynman hadn't taken much interest in the shuttle programme, having noticed that none of the results from the supposedly important scientific missions that it carried into Earth orbit were ever published in the main scientific journals, and suspecting that the whole exercise was a bit of a boondoggle.[6] But like millions of other Americans he saw the lift off and explosion of the *Challenger* on the TV news.

What Feynman didn't know was that the Acting Head of NASA was William Graham, who had been a Caltech undergraduate 30 years before and had attended Feynman's famous Physics X course. Graham had gone on to work at Hughes Aircraft, where he often attended Feynman's Wednesday lectures, sometimes accompanied by his wife. Graham had the unenviable task of drawing up a shortlist of candidates to participate in the inevitable Presidential Commission looking into the causes of the disaster. Most of the people who ended up on the Commission had some sort of expertise involving the space programme – which, unfortunately, meant that they could not truly be regarded as disinterested investigators, no matter how hard they might try to be impartial. They included Air Force General Donald Kutyna, who was responsible for shuttle operations for the Department of Defense, Sally Ride, the first American woman in space, Neil Armstrong, the first person on the Moon, and other people associated either with NASA or the space programme. The Chairman of the Commission would be William Rogers, a former Secretary of State and Attorney General. When Graham's wife suggested asking Feynman, a truly independent and original thinker, to join the team, Graham leaped at the idea.[7]

Graham phoned Feynman to ask if he would be available, not knowing that the request was particularly ill-timed. By now, Feynman had undergone two operations for abdominal cancer, was suffering from heart trouble and had been found to have another rare form of cancer, involving his bone marrow and affecting his blood, making it sticky and prone to clotting. Health concerns aside, he had spent much of his life avoiding responsibility, following his independent path and in particular steering clear of anything to do with Washington. His immediate reaction was to say no. But first he checked with his closest friends, including Gweneth. They all told him that he had to do it, because he could make a unique contribution. As Gweneth put it:

> If you don't do it, there will be twelve people, all in a group, going around from place to place together. But if you join the commission, there will be eleven people – all in a group, going around from place to place together – while the twelfth one runs round all over the place, checking all kinds of unusual things. There probably won't be anything, but if there is, you'll find it. There isn't anyone else who can do that like you can.

She was right.

So Feynman agreed to join the Rogers Commission, only to find that its remit extended far beyond finding the immediate cause of the disaster, to address questions such as 'What should be our future goals in space?' He foresaw the possibility that the Commission's work might be never-ending, and gave himself a deadline – he would serve for a maximum of six months, then quit, no matter what. But he would give Washington an honest six months, and wouldn't do anything else in that time – no teaching, no consultancy with the Thinking Machines Corporation, no physics. As he told Gweneth, 'I'm gonna commit suicide for six months.'

The call from Graham confirming that Feynman was a member of the Commission came at 4 p.m. on Monday, 3 February. He would be expected in Washington for the first meeting on Wednesday morning, which gave him a whole day to prepare, Feynman fashion, for the task ahead. He arranged with Al Hibbs, one of the friends who had urged him to serve on the Commission, to visit the Jet Propulsion Laboratory for an intense briefing on the shuttle, so that he would be up to speed. He learned a lot that day, but one of the most important things he learned was just about the first. On the second line of his handwritten notes from the briefing, he made the comment 'O rings show scorching'.

The O rings were part of the two solid fuel booster rockets that helped to launch the shuttle into orbit. The booster rockets are made in cylindrical sections and joined together. The O rings are like huge rubber bands, 37 feet in circumference, that fit into the joint between two sections of rocket and are supposed to seal the joint tight so that hot gas cannot escape through the crack as the fuel burns. After they have done their work, the spent booster rockets separate from the shuttle and fall into the sea, from where they are recovered and refurbished for future use. If the O rings on some of the recovered spent boosters were scorched, that meant that hot gas was escaping from the joint. If the seal failed entirely during a launch, it could cause just the kind of disaster that had engulfed *Challenger*. But why should the O rings have failed on the *Challenger* launch on 28 January 1986, and not on any of the previous shuttle launches?

Feynman 'sucked up information like a sponge' at JPL, without finding any answer to that question. Then he caught the overnight flight to Washington, and made it to the first meeting of the Commission, in Rogers' office, on Wednesday, 5 February. Wound up by the intense cramming of the previous day, and by lack of sleep, he was bothered to find that the first meeting was just an informal get-together, and nobody seemed to share his sense of urgency about getting to grips with the real work. On the other hand, he was relieved to learn that the investigations were expected to take no more than 120 days, less than his promised six months.

Although Feynman didn't know any of the other members of the Commission, he couldn't help noticing General Kutyna, resplendent in his uniform among the group of civilians and who Feynman happened to sit next to at the first meeting. For once, though, it turned out that there was a human being inside the uniform – Feynman was delighted to find that while many of the commissioners were met by fancy limousines after the meeting, Kutyna was heading off for the Metro.

> I thought, 'This guy, I'm gonna get along with him fine; he's dressed so fancy, but inside he's straight. He's not the kind of general who's looking for his driver and his special car; he goes back to the Pentagon by the Metro.' Right away, I liked him.

The feeling was mutual, and Kutyna took Feynman under his wing, showing him the way things worked in the Washington bureaucracy:

> Feynman had three things going for him. Number one, tremendous intellect, and that was well known around the world. Second, integrity, and this really came out in the commission. Third, he brought this driving desire to get to the bottom of any mystery. No matter where it took him, he was going to get there, and he was not deterred by any roadblocks in the way. He was a courageous guy, and he wasn't afraid to say what he meant.[8]

Kutyna was lucky to find out the kind of person Feynman was, and to establish such a good relationship with him so quickly, because by the end of the week the General would have a problem. He would be given a strong hint about what had caused the *Challenger* to explode, but the information would come from a sensitive source, a NASA astronaut who could get fired for telling tales out of school. The fact that this was a real possibility is itself, of course, an indictment of the way NASA was being run at the time, but Kutyna knew it wasn't just paranoia. There had been a previous occasion when one of the astronauts, an old friend of Kutyna, had passed him a document describing how safety procedures had been violated during manufacture of the shuttle segments. The astronaut was seen passing Kutyna the document, and promptly demoted.

Shortly after the Commission started its work, another astronaut told Kutyna about some sensitive information. The contractor involved had been testing O rings under conditions of extreme cold for at least six months prior to the accident. Clearly, there was some concern about what happened to the O rings when they got cold. This was potentially a key piece of information, since the fatal *Challenger* launch had been the first shuttle lift off to take place when the temperature was below freezing. If cold were implicated in the disaster, perhaps causing a failure of the O rings, then those data ought to be available to the Commission – but they had never been mentioned in the material that was to go before the commissioners. Kutyna desperately needed to get the possibility out in the open, without damaging the career prospects of his astronaut friend. The best way to do this would be if he could steer Feynman, the only truly independent member of the Commission, into thinking about the problem of cold affecting the O rings.

But he would have to be subtle. Right at the beginning of their work on the Commission, Kutyna had given Feynman a personal briefing, at the Pentagon, on the whole space programme, to put their deliberations on the shuttle in perspective. He had offered to get Feynman clearance to have classified information, but Dick had refused, saying, 'I don't want to clog my mind with secrets that I can't talk about. I want to be able to talk about anything that you tell me. So don't give me anything classified.'[9] So there was Kutyna's dilemma. The only person on the Commission he could trust to put the puzzle of how cold affected the O rings on the agenda was also the only person on the Commission who adamantly refused to be involved in secrecy.

Feynman was also the only person on the Commission who was unaccustomed to the slow pace of their work. After the informal meeting on Wednesday, lasting just a couple of hours, they were free for the rest of the day. On Thursday, in the first public meeting, the commissioners had the opportunity to question senior representatives from NASA. It turned out that all but a couple of the members of the Commission had degrees in science or engineering, and they fired off technical questions that the administrative 'big cheeses', as Feynman called them, were not equipped to answer. 'We'll get back to you on that' became the litany of the day. Friday wasn't much better. Although Kutyna gave the commissioners a rundown on the way an earlier investigation he had worked on, into the causes of a failure of an unmanned Titan rocket, had been carried out, Rogers (one of the few people on the Commission with no technical background) dismissed this practical experience as inappropriate to the shuttle investigation. 'We won't be able to use your methods here', he told Kutyna, 'because we can't get as much information as you had.'

To Feynman, this was patently false. Because the Titan was unmanned, it didn't have anywhere near as much monitoring equipment as the shuttle, and nor had the launch been filmed in close-up for national TV, whereas the pictures from the *Challenger* launch were good enough to show a flicker of flame coming from the side of one of the booster rockets just before the explosion. It was another frustrating day. 'Although it *looked* like we were doing something every day in Washington, we were, in reality, sitting around doing nothing most of the time.'

Then came the weekend. As it turned out, the Commission

faced a *long* weekend break. They were scheduled to go to Florida the following Thursday, to get a briefing from NASA officials and tour the Kennedy Space Center. Such a formal, guided tour had no prospect of providing any real insight into what went on; and even that cosmetic exercise was five days away! Devastated, and on the point of pulling out of the investigation, Feynman called Bill Graham, who had got him into it in the first place, and asked if there were any way he could get to do some real work, talking to engineers, trying to find out what had gone wrong. Graham thought this was a great idea, and offered to arrange for Feynman to visit the Johnson Space Center, as soon as he liked. But Rogers vetoed the proposal. Graham suggested a compromise – Feynman would stay in Washington, but Graham would arrange for NASA experts to give him a briefing at NASA headquarters, right across the street from Feynman's hotel. At first, Rogers objected to this proposal, too; but eventually gave it a reluctant OK.

So on Saturday Feynman began to get down to some real work on the problem, picking up where he had left off at JPL. When he talked to the expert who knew all about the seals in the joints where the pieces of booster rocket fitted together, it soon became clear that there was a known problem with the O rings, that had been brushed aside, largely (it seemed) through wishful thinking. There had been minor leaks on previous flights, and sometimes parts of the O rings on the recovered boosters had been burnt away. But only a few of the seals had failed, on only some of the flights. NASA's attitude, as Feynman described it, had been 'if one of the seals leaks a little and the flight is successful, the problem isn't so serious'. He likened it to playing Russian roulette. The first time you pull the trigger, the gun doesn't go off, so you assume it is safe to pull the trigger again, and again, and again . . .

He actually found a report which began 'the lack of a good secondary seal in the field joint is most critical' and concluded 'analysis of existing data indicates that it is safe to continue flying'. How could it be safe, if the situation were 'most critical'?

By now, the press were on to the story of the problems with the seals, and the following day, Sunday, a story appeared in the *New York Times*. One result of this was that Rogers called an emergency meeting of the Commission for Monday, 10 February. Kutyna phoned Feynman at his hotel on the Sunday afternoon, to inform him of the special meeting, and to invite him over to dinner that evening. He had heard the news about cold affecting the O rings almost a week before, and he was still looking for a way to put Feynman on the scent. He found it after dinner, when he was showing Feynman his pride and joy, an Opel GT 1974 that he was working on in the garage. There were a couple of carburettors on the workbench. One of the important components of such a carburettor is a seal formed by a rubber O ring, a miniature version of the shuttle O rings, to stop leaks where two of the subcomponents are joined.

'You know, Professor Feynman,' said Kutyna, 'these damn things leak when it gets cold. Do you suppose cold has any effect on the rubber O rings in the carburettor?'[10] It was enough to set Feynman on the trail to what would become his most famous, and public, experiment. Thanks to Kutyna's hint, he was already thinking about the effect of cold on the O rings when he went along to the special meeting of the Commission on Monday. The first part of the meeting was a waste of time. The 'exposé' in the newspapers hadn't contained any information that Feynman didn't know already. But then things got interesting. First, they were shown pictures they hadn't seen before, revealing puffs of smoke coming from one of the joints on the booster rocket before the shuttle had cleared the launch pad. The smoke seemed to be coming from the same place as the flame that had appeared just before the explosion, indicating strongly that the seals had been faulty and leaking from the very beginning of the launch.

Then came a real surprise. An engineer from the Thiokol company, responsible for the seals, spoke to the Commission. He had come on his own initiative, without being invited – if it hadn't been for the special meeting called because of the newspaper stories, he wouldn't have found the commissioners there. He told the meeting that the Thiokol engineers were so concerned about the possible effects of cold on the seals that the night before the launch they had advised NASA not to fly the shuttle if the temperature fell below 53° Fahrenheit, the lowest temperature at which the shuttle had flown before. But, said the engineer (Feynman only gives his name as 'Mr MacDonald'), NASA bullied Thiokol into reconsidering its opposition to the flight, which was launched, fatally, when the temperature was 29° Fahrenheit. Only MacDonald refused to go along. He told the Commission that he had said to his colleagues, 'If something goes wrong with this flight, I wouldn't want to stand up in front of a board of inquiry and say that I went ahead and told them to go ahead and fly this thing outside what it was qualified to do.' MacDonald's testimony was so stunning that Rogers asked him to repeat the whole story.

There were two aspects to MacDonald's story. First, it pinpointed cold as the immediate cause of the O ring failure. Kutyna's hint to Feynman had given him a bit less than 24 hours' start on the rest of the commissioners, but even without that hint, after MacDonald's testimony he would have been hot on the trail. Secondly, it showed, as Feynman had begun to suspect after his briefings on Saturday, that there had been two failures. One was a technical failure; the other was a human failure, a failure of management. Even when the engineers had expressed concern, the managers had pressed ahead.

The news was so important that Feynman wanted to get to grips right away with finding out how the properties of the rubber in the O rings were affected by cold. But Rogers decided to call another public meeting for the following day, Tuesday – not to air MacDonald's news, which he considered too sensitive to go public with just yet, but to go over the old material that had been in the *New York Times*. The idea was to go over much the same ground that they had covered in closed session on Monday afternoon (and which had been old news to Feynman even then!), but to do it in front of reporters and TV cameras. Feynman hated the thought of wasting more time, when he was now in a position to get some real information about what happened to the O rings under freezing conditions. But he was stuck in a hotel in Washington, distanced from the labs where the necessary experiments could be carried out. It was while eating dinner alone that night that he noticed the glass of ice water on the table, and said to himself, 'Damn it, *I* can find out about that rubber . . . I just have to *try* it! All I have to do is get a sample of the rubber.'

He knew that there was always ice water available at the Commission meetings, and thought about doing the experiment, for real, while they were all sitting around listening to the same old stuff they had heard already. The idea was irresistible to the showman side of his personality. But first, he needed a sample of the rubber used in the O rings. Once again, he called Graham, who came to the rescue. There was a model of one of the joints at NASA headquarters, which was going to be shown during the open meeting the next day. It contained two strips of the rubber (the O rings were only about as thick as an ordinary pencil, in spite of the importance of their job; what mattered was their flexibility, their ability to squeeze into the tiny gaps that opened up between the joints in the rocket under the stress of take-off, and block the exit of any hot gases). But Feynman would have to get the sample of rubber out of the joint himself.

Early next day, Tuesday 11 February, Feynman dropped by a hardware store and bought a few tools, including a small C-clamp. Then he went over to Graham's office. All he needed, it turned out, was a pair of pliers to pull the rubber out of the joint. There and then, he tried the experiment (for some reason, in *What Do You Care* Feynman said he was 'ashamed' of having 'cheated' by trying the experiment in private first; it seems like a sensible precaution to us!). Then, he put the rubber back into the model joint ready for Graham to present it to the meeting.

At the meeting, Feynman sat, pliers in one pocket and C-clamp in another, next to General Kutyna. Everything was set – except there was no ice water. Urgent requests produced ice water for everybody, not just Feynman, after the meeting had begun but, fortunately, before the model joint had been displayed. Kutyna was aware something was going on. As the joint was passed around, it came to him and he gave it to Feynman. A NASA spokesman was explaining how the seals worked, while the commissioners pretended they hadn't heard it all before. When the joint reached Feynman:

> He laid it in front of him, reached in his pocket, and got out a pair of pliers, a screwdriver, and a clamp. I thought, 'Oh my God, what's he going to do?'
>
> He proceeded to take this thing apart. He was going to take a piece of this O ring rubber, put his clamp on it to compress it, like it got compressed in the shuttle joint, then put it in ice water to cool it

down to the temperature on the day of the launch, and show that the O ring did not bounce back to its original form.[11]

Eager to show his experiment to the world, and relieved that the ice water had, after all, arrived in time, Feynman reached for the red button in front of him. Pressing this would indicate that he wanted to make a contribution, switch his microphone on and get the TV cameras and lights pointing his way. But Kutyna, watching what was going on, realized that the focus of attention was elsewhere. 'Not now', he told Feynman. It happened again. Kutyna told Feynman to wait. He flipped through his briefing book, and pointed out a particular diagram to Feynman. 'When he comes to this slide, here, that's the right time to do it.' The moment arrived, and all eyes turned to Feynman. He showed them his experiment and explained what was going on:

> I took this rubber from the model and put it in a clamp in ice water for a while . . . I discovered that when you undo the clamp, the rubber doesn't spring back. In other words, for more than a few seconds, there is no resilience in this particular material when it is at a temperature of 32 degrees. I believe that has some significance for our problem.

The demonstration didn't make the immediate impact Feynman had expected. His fellow commissioners seemed irritated by what they saw as clowning around, while the media representatives seemed puzzled by it, and the questions they asked Feynman during the lunch break were so mundane ('Would you explain to us what an O ring is, exactly?') that he thought they had missed the point, and grumpily blamed Kutyna for not letting him press the red button when he first wanted to. But that night, Feynman's experiment was on all the major TV networks (it was also shown around the world), and next day it was a major story in the *New York Times* and the *Washington Post*. Delighted, Feynman put his arm round the General, and said, 'Hey, Kutyna! You're not all bad!'[12]

> I don't think any of us could have done the experiment. It just would not have been fitting for a two-star general, or a former Secretary of State, or the first man on the moon, to pull out his beaker of water and do that kind of thing. But Feynman was able to do that. I guess if he had a weakness, it was for showmanship. He was a superb showman.[13]

He was also a superb scientist. *If it disagrees with experiment, then it is wrong.* Wishful thinking might say that the rubber would carry out its job when the temperature fell below freezing, but all it took to prove it wouldn't was a glass of ice water and a C-clamp. Any of the Thiokol engineers, had they thought in the way Feynman thought, could have done the experiment before the launch. But whether even that would have persuaded NASA to postpone the launch is debatable. *The easiest person to fool is yourself*, and the NASA bureaucracy had been fooling itself that all was well for far too long.

Feynman became a national hero and a public figure as a result of his little experiment, which had been carried out less than a week after he arrived in Washington. As Freeman Dyson has commented, it was his 'finest hour as a communicator', in which 'the public saw with their own eyes how science is done, how a great scientist thinks with his hands, how nature gives a clear answer when a scientist asks her a clear question'.[14] What the public didn't see was Feynman's continuing work with the Commission over the next few months, probing into the problems of management that had allowed the advice of the engineers to be overlooked, and had led to the death of seven astronauts.

This was, perhaps, the most important part of the Commission's work, and it was chiefly thanks to Feynman that it got done at all. As Al Hibbs has explained:

> By forcing this into the open, and doing it right there on television for the world to see, the rest of the commission could not avoid it any longer, and they had to say, 'Yeah, that's it. Now, why did it happen?' They might have spent all their time looking at what happened, considering all the technical possibilities, and never getting around to 'why?'
>
> I think that he prevented the complete bureaucratic whitewash that it might have turned out to be, saying, 'Nobody's really to blame, it was an unfortunate accident,' and so on. Feynman said, 'No, that's not true. Lots of people were to blame. The system was to blame. And you've got to say that. You've got to say it openly.'[15]

The investigation also uncovered other engineering problems, especially with the engines, that took years to correct before the shuttle flew again. Feynman carried out exactly the maverick role that Gweneth had known he would, cutting through the fog, seeking out real facts, even if it meant making himself a nuisance. And he did this with no regard for his own health or wellbeing. It was sadly apparent to all who knew him, when he returned to Caltech, how much it had taken out of him.[16]

The story of his struggle to get his own views into the official report of the Commission, and how they eventually appeared as an Appendix, not part of the main body of the report, has been detailed in *What Do You Care* (which also includes that Appendix to the report of the Rogers Commission). Popular accounts of this investigation often give the impression that Feynman was uniformly critical of NASA; in fact, although he was highly critical of the situation concerning the engines, he was happy with the situation on the avionics side, and positively glowing in his endorsement of the computer experts responsible for flight simulations: they were people who 'looked like they knew what they were doing' (*very* high praise, coming from Feynman). It was a genuinely balanced report, praising the good but not afraid to point the finger at the bad. And its final sentence is a characteristic piece of Feynman wisdom:

> For a successful technology, reality must take precedence over public relations, for Nature cannot be fooled.

As the last word on Feynman's last piece of technical work, that could not be bettered.

Notes

1. Carl Feynman, in *No Ordinary Genius.*
2. Conversation with Mehra, 1988.
3. See the contribution by Hillis to *Most of the Good Stuff.*
4. See note 3.
5. See note 3.
6. See *What Do You Care.* Half of that book is taken up by Feynman's own account of his experiences on the *Challenger* inquiry, and this is the primary source used, except where stated, in the rest of this chapter.
7. See Gleick.
8. Kutyna, in *No Ordinary Genius.*

9. See note 8.
10. See note 8.
11. See note 8.
12. He revised his opinion of the General again after the inquiry was over, when Kutyna made a point of confessing how he had deliberately pointed Feynman towards the possibility of cold affecting the O rings; but then Feynman forgave Kutyna. Neither of them, though, seems to have given full credit to MacDonald, thanks to whom Kutyna's little subterfuge wasn't really necessary.
13. See note 8.
14. Freeman Dyson, *From Eros to Gaia* (Pantheon, New York, 1992).
15. Hibbs, in *No Ordinary Genius.*
16. Helen Tuck, interviewed by JG in April 1995, simply said of the *Challenger* investigation 'he became very ill, during that'.

13

The final years

The quest for Tannu Tuva is one thread of adventure that runs through the last 10 years of Feynman's life; but, as we have seen, there were plenty of other things going on in his life during those 10 years, including his work for the Thinking Machines people, and the shuttle investigation. For most of the time, Tuva was well into the background of Feynman's activities, and the adventure was as much Ralph Leighton's adventure as it was Richard Feynman's. But it was always there, in the background, and it was typical both of Feynman's approach to life and of the way he passed on his enthusiasm to other people that, eventually, the wild scheme of organizing an expedition to visit a remote part of the Soviet Union, chiefly on the grounds that the name of its capital city contains none of the five ordinary vowels, did eventually reach fruition. The fact that this achievement was largely organized by a secondary school teacher who just happened to be a friend of Feynman also shows just how far any of us might go in achieving our wildest dreams, if we took on board a little of the Feynman spirit of adventure.

In fact, the Tuva adventure got off to a slow start. Although in January 1979 Radio Moscow made a programme about Tuva in response to a letter from Leighton asking for information about the region, it added little to what they had already gleaned from encyclopaedias and other reference books. But Ralph and Dick were delighted and encouraged by the closing comments of the narrator – that Tuva is now 'easy to reach' by airliner from Moscow.[1] Unfortunately, it later turned out that this meant it was easy for Soviet citizens to fly to Kyzyl, Tuva's tiny capital; it was not on the officially approved tourist trail for foreigners. In his enthusiasm, Leighton played a tape of the broadcast to his geography class the next day, not stopping to think of the possible repercussions on his career – schoolteachers weren't expected to play tapes from the programmes of Radio Moscow in class (the Soviet Union was still officially regarded, at that time, as the 'Evil Empire'). But in this case, nobody seems to have taken exception.

Later in 1979, they obtained a Tuvan–Mongolian–Russian phrasebook, and laboriously used it to compose a brief letter, in Tuvan, which they sent to the Tuvan Research Institute of Language, Literature and History, in Kyzyl, which had produced the book. By the time a reply came back, well into 1980, the Soviet Union had invaded Afghanistan, American–Soviet relations had sunk close to an all-time low, and the prospects of a couple of ordinary American citizens getting to visit a remote region of the Soviet Union were more slender than ever. But at least they were in contact with someone in Tuva!

Freeman Dyson has provided us with a delightful snapshot of life in the Feynman household at the end of 1979, in his book *From Eros to Gaia.* In a letter describing a visit to the West Coast, he says that 'the best thing that happened was a supper with Dick Feynman at his home in Pasadena'. This was the first time they had met for 12 years, and Dyson was delighted to find Feynman seemingly in much better health than rumour had led him to expect. 'He is still the same old Feynman that drove with me to Albuquerque 30 years ago', he wrote. 'Feynman has been married for about 20 years to an English wife called Gweneth. He enjoys the domestic life and they have a menagerie very much like ours, 1 horse (for the 12-year-old daughter), 2 dogs, 1 cat, 5 rabbits. But they have temporarily outdone us, for the next few months, by taking on a boa-constrictor who belongs to some neighbours on a leave of absence.'

As it had been for the previous 15 years, the other consuming passion in Feynman's life, after physics and his family, was still drawing. He drew or painted every Monday evening, getting better in a mysteriously erratic fashion, with occasional jewels interspersed with less successful efforts – some of his work, together with the background to Feynman's interest in art, has now been published as *The Art of Richard P. Feynman.*

With Feynman so happy at home and busy with Thinking Machines, drawing and other activities, at first the Tuva project was no more than a pipe dream. By 1981, after three years of intermittent discussion of the scheme, he and Leighton were no closer to Tannu Tuva. Then, in the autumn of 1981, Feynman's cancer struck again. This particular kind of cancer doesn't jump from one part of the body to another, leaping from the kidney to, say, the lungs, but spreads out more slowly from its original site. In this case, the cancerous tissue was now wrapped around Feynman's intestines. Once again, immediate drastic surgery was the only hope of holding its advance at bay.[2]

To Feynman, his illness was as much an adventure as anything else that happened to him. He described it, in his characteristic way, as 'int-er-es-ting' (he always gave all four syllables of the word their full weight), and studied it the way he would a problem in physics. In this, he seems to have shared his father's ability to look at his own illness from outside, as it were. Joan Feynman has recalled how Melville, who knew he suffered from dangerously high blood pressure, once said, 'Have you seen my bloodshot eye? Now, that's an interesting thing, because . . .', ending his explanation of why the blood vessels in his eye were damaged with 'One day, that's going to happen in my brain.'[3] At the time of his first operation for cancer, Richard instructed the surgeon that if it looked likely that he wasn't going to recover, he wanted to be brought out of the anaesthetic, so that he could 'see what it was like to go out'. He felt that it would be cheating to die under anaesthetic. If he was going to die, he wanted to see what it was like.[4]

The second cancer operation, in which another large chunk of Feynman's insides was removed, lasted for more than 10 hours, and did not go smoothly. An artery near his heart ruptured, and he suffered a massive loss of blood. By chance, two other patients with the same blood group as Feynman (type O) had also needed major transfusions that day, and the blood bank at the UCLA hospital was running low. An emergency call went out, producing a line of 100 volunteers, mainly students and staff from Caltech and JPL, giving their blood to keep Feynman alive. Altogether, he needed nearly 80 pints before the emergency was over.

Even Feynman couldn't bounce straight back from an ordeal like that. But he had an incentive to get back on his feet and a target to aim for. In 1982, the yearly Caltech musical production was to be

South Pacific, and Feynman and Leighton were asked to provide the drumming for a scene with Tahitian-style dancers. They took lessons from a drummer in Los Angeles who knew Tahiti well (as Feynman was fond of saying, you can find *anything* in Los Angeles), and Dick even learned a few phrases of Tahitian to shout out during the performance.

For the show, which took place barely three months after Feynman's second cancer operation, he was dressed as a tribal chieftain, with a tall headdress of feathers and a long cape. It was during the rehearsals for *South Pacific* that Leighton, taking his cue from the director of the show, took to referring to Feynman as 'the Chief', a name which stuck for the rest of his life. On opening night, Feynman, still weak, had to sleep through most of the show, getting up only to make his contribution on stage. But in his brief appearance he looked to the audience to be fully recovered and in prime form. It was his first public appearance since the operation, and many of the people who had given blood to keep him going through that ordeal were in the audience; hardly surprisingly, the cameo appearance was the highlight of the show, producing a standing ovation.

It is also hardly surprising that after his second operation Feynman began to take more interest in joining in some activities that others might have regarded as dippy, but which actually echoed his long-standing interest in how the mind works, dating back to his student days. Through his lectures at Hughes, as he describes in *Surely You're Joking*, Feynman had met John Lilly, who was carrying out experiments in sensory deprivation, designed to produce hallucinations as the subject floats in a tank of water at body temperature, completely in the dark. Feynman eagerly tried the tanks out, following up his fascination with what happens to the mind as you go to sleep by trying to produce hallucinations while still awake. He succeeded, but never found anything in his 'out of the body' experiences to persuade him that it was anything other than an hallucination, produced entirely by the internal workings of the brain, rather than a genuine view of his body from the outside. He also found that although it would usually take about 15 minutes to produce the hallucination, he could do so much more quickly if he smoked a little marijuana beforehand – the physicist who had given up drinking alcohol because he didn't want to damage his thinking ability was now prepared, cautiously, to try hallucinogens in his quest to find out how the brain worked, aware that he was living on borrowed time and wouldn't be using that brain for much longer, anyway.

In some ways, Feynman's zest for life increased after his second operation, as he determined to make the most of whatever time was left to him. He became an annual visitor at the Esalen Institute at Big Sur, south of Monterey, a kind of hippie centre for many 'alternative' or 'holistic' ideas. At Esalen, there are some large baths fed by hot springs, situated on a ledge about 30 feet above the Pacific Ocean. What Feynman described as 'one of my most pleasurable experiences' was to sit in those baths watching the waves crashing on to the rocky shore below, and gazing into the clear blue sky above.[5] In return for the pleasure he got out of visiting Esalen (where he also learned the art of massage), he gave talks on 'Idiosyncratic Thinking', and told the assembled new-agers about tiny machines and quantum mechanics. 'The Chief never forgot he was living on borrowed time', says Leighton, who recalls relaxing in the baths with Feynman in the mid-1980s and hearing him suddenly cry out 'Thank you, Dr Morton!' Dr Morton was the surgeon who had held the cancer at bay, and Feynman would thank him for his extra years of life 'in the same way that others would thank God for giving them another fine day'.[6]

That moment still lay ahead in the spring of 1982. The Tuvan scheme showed no signs of getting off the ground, and the world seemed to be in a mess, with Argentina invading the Falkland Islands and Israel invading southern Lebanon. To alleviate the gloom, in June Leighton took Feynman to Las Vegas for a belated sixty-fourth birthday present. To their delight, the hotel provided complimentary 'funbooks' of coupons that could be used to make bets. To their even greater delight, after betting with all the vouchers in their books they had made a profit of some $50, and were careful to make no further bets. But as they were leaving the hotel, a couple of days later, they found that their key deposit could be returned to them in the form of funbooks, instead of cash. Off they went to the gaming tables, and started winning again – only to be requested, in no uncertain terms, to leave the table. There were still some vouchers left in the funbooks – but they could proudly boast to their friends when they got home that they had been sent away from the tables in Las Vegas for winning too much.[7]

It was an adventure just like the ones that Feynman told Leighton about during their drumming sessions. Not long after, Leighton went to Esalen with Feynman for the first time, to teach drumming as a kind of antidote to the heavy physics involved in Feynman's lectures on 'The Quantum Mechanical View of Reality'. The material covered some of the same ground that he was to cover early in 1983 at UCLA as the Alix G. Mautner Memorial Lectures, and which was turned into a book by Ralph Leighton (continuing the family practice which his father, Robert Leighton, described as 'translating lectures from Feynmanese into English').

Those lectures came about through Feynman's lifetime friendship with Leonard Mautner, one of his boyhood companions and a fellow mathematics enthusiast from Far Rockaway. Like Feynman, Mautner ended up on the West Coast, but in his case at UCLA; Mautner's wife, Alix, was a specialist in English literature, but had a keen interest in science, and often asked Feynman to explain things to her, in a friendship lasting more than 20 years. But he never had time to get to grips fully with an explanation of quantum electrodynamics for her, and promised that one day he would prepare a series of popular lectures on the subject, that she could attend.[8] Eventually, he had an opportunity to prepare just such a set of lectures and, as he put it, 'try them out', when he was invited to visit New Zealand at the end of the 1970s. He gave a variation on the theme on a visit to Crete in the early 1980s, as well as using the material at Esalen, polishing his performance all the time. The lectures went well, but in 1982, before he could put on the definitive performance in Los Angeles for his friend, Alix died. So the lectures on QED at UCLA in 1983 became the first of the Alix G. Mautner Memorial Lectures.

They were the ultimate Feynman lectures – the master himself, at the height of his powers as a showman, explaining in simple, everyday language the work for which he had won the Nobel Prize, and which remains the jewel in the crown of theoretical physics. The kind of showman (or shaman!) Feynman was was explained in an obituary that appeared in *Scientific American* in June 1988:

> The actor on the stage pretends to be who he is not, by artful empathy and the words of another. That was not Richard's way. His theater – and it is impossible to evoke him without the word 'theatrical' – was on the other side. Richard's was the stage where dancers, wire walkers and magicians daringly perform. What they do is striking, and not dissembled or illusory. It is real, expressing mastery of some challenge, trivial or urgent, posed by nature and by human perceptions. On that stage he performed in four real dimensions.[9]

Nowhere was this mastery of a challenge more evident than in those lectures on QED. The resulting book, *QED: The Strange Theory of Light and Matter*, is a masterpiece of clarity, even though it pulls no punches and describes QED the way it really is, without sacrificing accuracy for simplicity. Leighton's role as Feynman's scribe was established by the success of the collaboration on *QED*, but by the time the book was published (by Princeton University Press in 1985) it had been overtaken by *Surely You're Joking*, which Leighton had worked up from the tapes of his drumming sessions with Feynman in 1984, and which was also published (by Norton) in 1985.

The publishers demonstrated little real enthusiasm for the book, offering an advance of just $1500, and printing only a modest number of copies for the first edition. They were astonished when it became a bestseller. A few of Feynman's colleagues were disappointed by the seemingly frivolous tone of his anecdotes, even though he was always careful to say that that was all the book was – a set of anecdotes, not an autobiography. But thousands of people who never knew that physics could be fun were excited and intrigued by the book, and many of Feynman's old friends recognized the truth underlying his colourful stories – the truth about not fooling yourself, and always being honest. Someone who knew Feynman well enough to be a good judge of the honesty of the book, and its sequel *What Do You Care* (published after Feynman's death) is Freeman Dyson, who described them as providing 'a complete picture of Feynman in his own words'. Referring to the occasion when they shared a room in a brothel together and discussed life and physics, Dyson comments 'his version is different from the version I wrote to my parents. In deference to my parents' Victorian sensibilities, I left out the best part of the story. Feynman's version is better.'[10] Feynman, of course, would never leave out the best part of a story for the benefit of anybody's sensibilities, which is why his anecdotes ruffled a few feathers. And Dyson puts his finger on a fundamental reason for Feynman's integrity and sometimes uncomfortable insistence on always calling a spade a spade – 'Arline's spirit stayed with him all his life and helped to make him what he was, a great scientist and a great human being.'

Ralph Leighton summed up Feynman's approach to storytelling in 1995:

> Feynman would relate a story which would require several re-tellings to get it right. I don't think he would change a fact or invent something that didn't happen. But I know that he engaged the reaction of the listener, and developed the story, the impact, the effectiveness, as a good storyteller does. It just happens that the material of his stories involved himself. But most importantly, they had a kind of point to them . . . I think he went about them in the way the Dalai Lama does, and other great teachers, which is to teach you when you don't even realize you're being taught – through humour. Underlying all this is a philosophy, that it's good to have different points of view, to have a surprise at something being not the way you thought, to have authoritative figures make buffoons of themselves, so you will not be afraid of them, but can stand up to them; and not to believe what someone says just because he's wearing a uniform, or whatever.[11]

Feynman received an enormous amount of fan mail as a result of the book, all of which was opened and read by Helen Tuck. Feynman himself was too busy, and soon too ill, to handle much of it personally, although she was careful to pass on anything that needed a response. She recalls[12] that out of 'boxes and boxes' of mail there was only one letter expressing dissatisfaction with the book. It came from 'an old lady, bless her heart, in Long Beach . . . I think that was the only letter that came in that was really unhappy, and she was sorry she'd spent the money. So he actually sent her a cheque for the money, and he wrote her a nice letter.'

Leighton's tapes include many conversations that never made it into the books. In one, Feynman discusses his medical condition. He had been over to the Huntington Medical Library, to read up on the kidney – he only had one, now, and it was beginning to cause problems. 'It's all interesting, how the kidney works, and everything else', he said. 'You want me to tell you some interesting things? The damn kidney is the craziest thing in the world!'[13] In fact, Feynman became so absorbed by how the kidney works that the library closed before he got on to reading about his own particular problem, and he had to return another day for that.

Cancer and potential kidney failure were not Feynman's only medical problems. Like his father, he had high blood pressure; he also suffered from hypoglycaemia and a recurring arrhythmia of the heart. One attack of arrhythmia occurred when he was at Esalen with Ralph. Feynman called his doctor back in Pasadena, who said that although Feynman wasn't in any immediate danger, he should get back to Pasadena at once for a check-up. Before they left, someone who Ralph describes as a 'hippie doctor' at Esalen prescribed his own course of treatment, urging Feynman to drink a large amount of fizzy pop, which he did. Ralph and Dick had driven only a little way down the road when Feynman produced a large burp, and his heartbeat settled down into its normal pattern. He happily abandoned the trip back to Pasadena, and they returned to Esalen, to the delight of the hippie doctor, who was able to tell everyone how effective his treatment had been, with no recourse to drugs.

On another occasion, a major medical problem was, in a sense, self-inflicted. Feynman had gone downtown to collect one of the first IBM personal computers, jumped out of his car and stumbled across the sidewalk, hitting his head on the wall of the building. He cut his head severely enough to go to the hospital for stitches, but otherwise seemed OK. Over the next few weeks, though, he started behaving strangely.[14] He wandered about in the middle of the night for no good reason, and once spent three-quarters of an hour looking for his car, which was parked right outside the house. After three weeks, the problem reached crisis level, when Feynman was giving a lecture at Caltech and suddenly realized that he was talking complete nonsense (and nobody in the audience had had the courage to tell him, as he would have done if the situation had been reversed). He apologized to the audience, and went off to the hospital, where a brain scan showed that slow bleeding inside his skull had led to a build-up of pressure affecting his brain. The remedy was simple – two holes drilled into his skull let the fluid out and relieved the pressure on his brain. Next day, he was sitting up in bed, mentally alert, completely his old self – except that he had no memory of the three weeks that had passed since the accident. He greatly enjoyed telling friends, 'feel here; I really have got holes in my head!'

That autumn, though, Feynman was able to revive his oldest, and one of his closest, personal relationships. His sister Joan had spent most of her life in the eastern United States, where she had married, had had children, and a career. By the beginning of 1984 the last of her children had left home and she was living alone. In *Most of the Good Stuff*, she recalls how one February day in 1984 she was looking out of the window at the falling snow:

> When the thought came to me 'What am I doing here? Where would I rather be?' Richard already had cancer and I realized that if I ever was going to spend more time with him it had better be soon. So I called some friends at the Jet Propulsion Laboratory in Pasadena and told them I wanted to come out. I was lucky and the next fall I joined the lab and rekindled the relationship with Richard.

She found that he hadn't changed much. Although older and more famous (at least among scientists), he was just as excited with life and science as he had always been, and still as ready to laugh:

> All his life, he had done physics for fun and he was still doing it for fun. He said that when people asked him how long he worked each week, he really couldn't say, because he never knew when he was working and when he was playing.

Joan became part of the Feynmans' domestic scene in Pasadena, visiting the family for a meal every Thursday night, and spending long hours talking with her brother, or going for long weekend walks with him.

1984 had been a year of several failed schemes in the quest for Tuva. The following year, heady with the success of *Surely You're Joking* and *QED*, Leighton decided to travel, with a Russian-speaking friend Glen Cowan, to the Soviet Union, to see first hand the obstacles that he and Feynman were facing in trying to reach Tuva.

This was still in the 'Evil Empire' days, but the pair enjoyed a series of Feynmanesque adventures, recounted in *Tuva or Bust!* At the end of their trip, they made contact, in Moscow, with Sevyan Vainshtein, the author of one of the books they had acquired about Tannu Tuva, and with whom Feynman had been in correspondence. Vainshtein had what is surely the ultimate Feynman anecdote: when Vainshtein had been travelling in a remote part of western Tuva he once met a young woman, sitting outside a yurt (the traditional tented home of the nomadic people of the region), reading a book. She was studying to be a teacher, and the book she was reading was *The Feynman Lectures on Physics.*

The Russian translation of the *Lectures* had, it turned out, been the biggest success ever of the Mir publishing house, with more than a million copies sold over the previous 20 years. Feynman, of course, received no income from this, which was essentially a pirated edition; but this was no loss, since he never received any royalties from the original editions and official translations, either. Since the lectures had been given as part of his duties at Caltech, all the income from the books went to Caltech itself – a not entirely unreasonable arrangement, since for more than 20 years Feynman was, as we have mentioned, the highest paid member of the Caltech faculty. Not that he cared a fig about the pay, as long as he had enough to live on. What was much more important to him was that he was exempted from serving on faculty committees and the like – that he really did not have any 'responsible position'.

Vainshtein was an ethnographer, and Leighton and Cowan learned from him about an exhibition called 'On the Silk Road', which had been to Japan in 1982 and Finland earlier in 1985, exhibiting artefacts associated with the people that lived near the ancient Silk Road between Europe and China. Many of the pieces came from Tuva, and some of them had been found by Vainshtein himself. The exhibition would be going to Sweden in 1986. Leighton realized that he had been handed the perfect opportunity to make the Tuva dream come true. 'After Sweden,' he told his host, in one of the many obligatory toasts of vodka, 'the exhibition will come to the United States – and as members of the host museum, Richard Feynman, Ralph Leighton and Glen Cowan will visit Tuva with Sevyan Vainshtein!'[15]

In the summer of 1985, Feynman made his own last major trip abroad, to Japan. There had been a longstanding invitation for him to visit the University of Tokyo, but illness had prevented him from taking up the invitation before. Now, he was to be one of the chairmen of a conference held to mark the fiftieth anniversary of the work by Hideki Yukawa which predicted the existence of the family of particles known as mesons. It was largely an honorary job, since there would be two Japanese-speaking co-chairmen to ensure that things actually ran smoothly, but it was a great excuse for Richard to travel to Japan with Gweneth. They travelled widely, staying for some time in a tiny Japanese-style inn in the countryside, with no concessions to Western habits, just the way they liked it. After a perfect holiday, they returned to California at the end of August.

Nothing happened on the Tuva front until February 1986, when Leighton decided, on the spur of the moment, to go to Sweden to check out the Silk Road exhibition. Glen Cowan agreed to come along, but Feynman had just accepted the invitation to serve on the inquiry into the *Challenger* disaster, and was not available. While Feynman was finding out about the effect of cold on the shuttle O rings, Leighton was making contact with the exhibition organizers, and learning about the bureaucratic hoops they would have to jump through at the Soviet Academy of Sciences in order to get the exhibition to the United States. The main thing the exhibition people were interested in was ensuring that a large number of Soviet delegates would get a chance to travel to America with the exhibits; the way at last seemed clear for Feynman to visit Tuva.

If only Feynman had been there to share the pleasure of making the breakthrough, the happiness of Leighton and Cowan would have been complete. Then, that evening, they switched on the local Swedish TV to watch the news. 'Suddenly, there was the Chief, holding a little C-clamp in his hand, explaining something. For us, it was the icing on the cake; the third musketeer of the Tuva trio had suddenly appeared in Sweden after all.'[16]

On his return to Los Angeles, Leighton went to the Natural History Museum to see if they would be willing to host the exhibition, taking along with him some catalogues from the Swedish version of the show. The museum representatives were cautiously interested, but asked what the participation fee to the Soviet Academy of Sciences would be. There isn't one, Leighton explained – apart from the cost of hosting 14 Soviet representatives and taking them to Disneyland. The museum people were a bit more interested. And what about a finder's fee for Leighton and his colleagues? 'Nothing', he replied. He explained their burning desire to get to Tannu Tuva. He was among fellow enthusiasts for exotic places, who immediately understood. There were no problems; the museum's director soon approved the project.

Feynman finished his work on the shuttle Commission in June 1986, and returned looking 'tired and haggard', in Leighton's words. But with everything fixed up at the American end, the whole Tuva project now hinged upon getting the required protocol through friends at the Soviet Academy of Sciences. In September, Feynman was best man at Ralph's marriage to Phoebe Kwan; a week later, Dr Morton was performing another major operation. On their return from honeymoon, Ralph and Phoebe visited Feynman at the UCLA Medical Center, where he was recuperating. While they were there, two representatives from the Natural History Museum also came by, to bring Feynman up to date on progress. Everything was agreed. The exhibition would be coming to Los Angeles in January 1989 – and the protocol specifically included a provision for representatives of the American side to go out to Tuva in the summer of 1988 to make a film of the sites where the artefacts had been found, to accompany the exhibition. Feynman was delighted. Once again, he was being recognized as an expert in something he was not supposed to know about. 'You see, man?', he told his friend, 'We're *professionals*. We're finders of international exhibitions!'[17] Together with Ralph and Glen, he was officially enrolled as a Research Associate of the Natural History Museum of Los Angeles.

Even then, it wasn't all plain sailing. Recovering from his third major operation took time, and Feynman would go on increasingly long walks with Leighton to build up his strength in readiness for the trip to Tuva. Meanwhile, the proposed filming trip seemed to be falling through. The proposal for the Tuva expedition was passed from the Soviet Academy of Sciences to the Ministry of Culture, which had no stake in the exhibition coming to America. As the negotiations with Sovinfilm dragged on through the summer, Feynman's best chance to reach Tuva slipped away.

In September 1987, a Soviet delegation came to Los Angeles to coordinate the planning of the exhibition. It was headed by Andrei Kapitsa, the person in charge of exhibitions for the Soviet Academy of Sciences. Kapitsa provided a direct link with one of Feynman's major contributions to science – he was the son of Pyotr Kapitsa, who had won the Nobel Prize in 1978 for his work in the 1930s and 1940s on low-temperature physics with liquid helium II. It was Pyotr Kapitsa who had actually coined the term 'superfluid' to describe the behaviour of liquid helium at very low temperatures. Richard and Gweneth entertained the delegation of three at their home, and Feynman and Andrei Kapitsa got on well, even though Feynman was uncomfortable at the prospect of linking an official visit to Moscow with his cherished trip to Tuva.

The exhibition planners (including the Feynmans) spent one day on a visit to Catalina Island, 25 miles from Los Angeles. A longer trip to Yosemite National Park was next on the agenda, but Feynman, tired after the outing to Catalina (which involved a boat ride lasting an hour and a half each way, through choppy seas) decided not to go, and Gweneth stayed with him.

It turned out that it wasn't just the fatigue of the boat trip and the busy schedule of events surrounding the workshop that had got to Feynman. His cancer had struck again, and in October 1987, just a year after his previous operation, he was back at the UCLA Medical Center being operated on for the fourth time for this problem. After this operation, literally almost half of Feynman's insides had been removed. Astonishingly (and partly due to having an epidural anaesthetic to aid his recovery), within weeks Feynman was back teaching a graduate course in quantum chromodynamics at Caltech, and although now often tired and clearly in pain, he again started the daily walks to build up his strength for the trip to Tuva.

By now, the scientific establishment in Moscow had got wind of Feynman's proposed trip, from Andrei Kapitsa, and was eager to have him visit them while he was in the Soviet Union. This came close to being an invitation of the very kind that he had wanted to avoid, a trip for Feynman the physicist with Tannu Tuva thrown in as a sweetener, not a trip for Feynman the finder of international exhibitions. But still, even if the Soviet Academy of Sciences did finally want to get involved, the deal had been initiated through the museum connections, and Feynman must have been well aware that it would be his last chance to make the journey. So the 'three musketeers' agreed to take up the offer from the Academy of Sciences, if it ever materialized, and work towards a trip to Tuva in May or June of 1988.

In November, Feynman made his last public appearance, which has been movingly described by the physicist John Rigden, in *Most of the Good Stuff*. Feynman had agreed to serve on a panel discussing 'What High School Physics Should Include' at a public meeting in Los Angeles on 14 November. In October, it looked as if he would be too ill to make it, and Rigden was asked if he would be willing to 'fill in for Feynman' and agreed to join the panel. On 12 November, he heard that Feynman now felt well enough to participate. Rigden offered to step down, but the organizers said there was no need, and they would simply have both of them on the panel.

The meeting took place in the auditorium of La Cañada High School, where Rigden met Feynman for the first time since 1983. He was shocked by Feynman's frail appearance, but impressed by his thoughtful responses to the questions posed for the panellists. But the most telling aspect of Rigden's memoir is his description of what happened after the formal part of the meeting, when people crowded around Feynman asking him questions:

> As I watched, I realized I was witnessing something extraordinary. Feynman's energies grew as he responded to question after question. The outside corners of his eyes were creased by the smiles that played over his face as he talked about physics. His hands and arms cut through the air with increasing vigor as their motions served to complement, even demonstrate, *his* explanations . . . It was the enjoyment he exuded as he stood there talking physics with an eager, receptive group of physics teachers that moved me. It was an enjoyment I could feel. When the session ended and Feynman, along with David Goodstein, walked out of the La Cañada High School Auditorium, I had the feeling that I was standing on holy ground.

This was Feynman the showman physicist in his element. And the same phenomenon, of a failing body being lit up from within by the enthusiasm of the man, was seen again a couple of months later, at the end of January 1988, when Christopher Sykes came to Pasadena to interview Feynman for a BBC TV programme about Tannu Tuva. As anyone who has seen that programme will know, the enthusiasm for physics, for adventure and for life was still there.

Just before that interview was recorded, Feynman had received another visitor eager to talk to him about his life and science. Jagdish Mehra is a physicist who became fascinated with the history of his subject, especially the birth of quantum mechanics, and has written several scholarly books on the theme. He had got to know Feynman in 1962, and as early as 1980 he had asked Richard's permission to write a serious scientific biography of him. They had met intermittently since then, with Mehra asking questions about various aspects of Feynman's life and scientific work. In December 1987, he called Feynman and suggested another visit, to finish his preparations for the book. Feynman's initial response was 'I don't think I want to go over the past again; I am too tired and depressed.'[18] But on 23 December Feynman called back, in a more cheerful mood, to say that Mehra was welcome to come and talk. 'Thanks very much for calling me', Mehra replied. 'I was thinking of coming early in March, would that be all right?' Feynman said, 'I don't know. It might be too late then.'

Worried by this comment, Mehra (who was based in Houston) changed his schedule, and went out to Pasadena on 9 January. The next day, he met up with Feynman, who agreed to take part in taped conversations with Mehra every morning at 10 a.m., except for Tuesdays and Thursdays, when he was teaching his class on quantum chromodynamics. In exchange, Mehra had to entertain Feynman with stories over lunch. Like Ralph Leighton a decade earlier, Mehra was in the right place at the right time, when Feynman, aware now that he hadn't long to live, wanted to talk about his life and work to a wide audience. Although often obviously in pain, thin and weak, according to Mehra Feynman clearly enjoyed their discussions, and was in top storytelling form. The interviews continued until 27 January. As well as discussing science, covering the ground we have also covered in this book (and which he had covered in earlier interviews, most notably with Charles Weiner, of MIT, for the American Institute of Physics Archive), Feynman talked about life, the quest for Tannu Tuva, love and the happiness of his marriage to Gweneth and his delight in his two children. After the last interview, Mehra drove Feynman back to his house, and made his farewells. As Mehra left, he knew that he had seen the last of 'a great physicist and a most extraordinary man'.

On 1 February, Christopher Sykes completed filming what was to be Feynman's last interview, which took place just after Feynman gave what turned out to be his last quantum chromodynamics class. Two days later, new tests were carried out on Feynman's failing body: his remaining kidney was failing, and the cancer was back. His life could have been prolonged by dialysis, but the returning cancer would have brought a painful death in a matter of weeks or months. Feynman preferred to accept the inevitable at once, provided the people closest to him could take it. He told Gweneth, who spoke to Joan on the phone, telling her, 'Richard says he wants to die, and that it's your decision.'[19] The two women agreed that it would be senseless to prolong Richard's suffering, and went to visit him together, at the UCLA Medical Center.

> When I came in he was lying there and he said: 'Decision?' Because he couldn't talk very well. I said: 'Yes, you're going to die.' And his whole body just relaxed.

For the few days that remained, Feynman was watched over by Gweneth, Joan and his cousin Frances, who had shared the house in Far Rockaway. Before he slipped into the inevitable coma resulting from kidney failure he apologized to Dr Morton for dying on him. But even after he was in a coma, things happened that Joan is keen that people should know about:

> In the coma, his hand was moving, and Gweneth said that the doctor had told her that the motion is automatic, and it doesn't mean anything. So this man who'd been in a coma for a day and a half or something, and hadn't moved, picks up his hands, and goes like this, like a magician, as if to say 'Nothing up my sleeve,' and then he put his hands behind his head. It was to tell us that when you're in a coma you can hear, and you can think.[20]

The other message which Joan is sure Richard wanted communicated came soon after that incident. He came out of the coma briefly, and said, 'This dying is boring.' Then he went back into the coma. Those were his last words; Richard Feynman died at 10.34 p.m. on 15 February 1988.

Early in March, a letter addressed to Feynman arrived from Moscow. Dated 19 February, it was the formal invitation to visit Tannu Tuva. When the Soviet Academy of Sciences learned of Feynman's death, nothing more was heard from them about the possibility of the other 'musketeers' making the trip. Nothing daunted, Ralph and Phoebe Leighton managed to make their way to Novosibirsk, in the summer of 1988, as guests of Vladimir Lamin, a historian involved with the Silk Road exhibition. And through Lamin's good efforts, they made it all the way to Kyzyl – not as hangers-on to a party riding on Feynman's fame as a physicist, but in their capacity as finders of international exhibitions, just as Dick himself would have wished. The exhibition did indeed come to Los Angeles in February 1989. 'It turns out', Leighton (who is now an Honorary Consul of the Republic of Tuva) recalls with justified pride, 'that we inadvertently brought over the largest exhibition of artefacts ever brought in to the USA, all through trying to get to Tuva.' And, of course, through living life the way Feynman lived his life. In June 1989, Gweneth Feynman, Glen Cowan and others were invited to visit Tuva privately in 1990. But on 31 December 1989 Gweneth died of cancer.

Richard Feynman provided his own best epitaph, in a conversation he had with Danny Hillis when they were out walking in the hills behind Feynman's house, not long after one of his operations. It was the moment when Hillis realized that the problem was really serious, and that Feynman was probably going to die soon. Noticing his subdued state, Feynman asked him what was the matter. Hillis told him that he was sad because Feynman was going to die – such straightforward honesty seemed natural in Feynman's company:

> Richard said, 'Yeah, that bugs me too, sometimes.'
>
> But then he said something which I wish I could remember exactly. It was to the effect of 'Yeah, it bugs me, but it doesn't bug me as much as you think it would, because I feel like I've told enough stories to other people, and enough of me is inside their minds. I've kind of spread me around all over the place. So I'm probably not going to go away completely when I'm dead!'[21]

There is indeed a little bit of Richard Feynman in all of us who have heard, or read, his stories; and we are all the better for it.

Notes

1. Ralph Leighton, *Tuva or Bust!* (hereafter referred to as *Tuva*).
2. Feynman's mother, Lucille, died a few days before this second major cancer operation. In the words of Ralph Leighton (letter to JG), 'she died peacefully, in her favorite chair, after eating dinner'. She was 86 years old.
3. Joan Feynman, in *No Ordinary Genius.*
4. Ralph Leighton, interview with JG, April 1995.
5. *Surely You're Joking.*
6. *Tuva.*
7. See note 4.
8. Richard Feynman, introduction to *QED.*
9. The obituary was simply signed 'an old friend'. The old friend was Philip Morrison.
10. Dyson, *From Eros to Gaia.*
11. Leighton, interview with JG, April 1995.
12. Interview with JG, April 1995.
13. Leighton, interview with JG, April 1995; see also Gleick.
14. Gweneth Feynman, as told to Gleick.
15. *Tuva.*
16. *Tuva.*
17. *Tuva.*
18. This account of Mehra's last encounter with Feynman is taken from the transcript (provided by Mehra) of a talk he gave at Cornell University on 24 February 1988, while the events were still fresh in his mind; the introduction to his book *The Beat of a Different Drum* is based on the same talk.
19. Joan Feynman, in *No Ordinary Genius.*
20. See note 19.
21. Hillis, in *No Ordinary Genius.*

— 14 —

Physics after Feynman

There is no clear distinction between physics after Feynman and physics before Feynman, not least because Feynman's own methods and way of thinking have become an integral part of research at the cutting edge in modern physics. Indeed, as we shall see, some of the most intriguing new developments in theoretical physics have come, not from any breakthrough into new territory beyond the scope of Feynman's work, but rather from taking old ideas of his that were far ahead of their time and incorporating them into modern physics in a new way.

The most striking example of this link between what Feynman was doing decades ago and what young researchers are doing today comes from the aspect of his work that was least sung during his lifetime, the study of gravity. As we saw earlier, this culminated in a course of graduate lectures that he gave at Caltech in 1962–63, alongside the second year of the famous undergraduate lectures. During that remarkable year, Feynman gave his first sophomore lecture every Monday morning, a gravity lecture Monday afternoon, and followed these up later in the week with his second sophomore lecture and his regular talk at Hughes. At most, 15 people attended each of the gravity lectures – but they included two students, James Bardeen and James Hartle, who went on to make major contributions to the development of the theory of gravity. This highlights the way in which Feynman, as a teacher, provided an inspiring influence on students who were one step distanced from him, even if his compulsion to solve every problem he came across himself made him sometimes a less than ideal thesis supervisor. Since Bardeen actually was one of Feynman's PhD students, however, it is certainly far from true that none of the students Feynman supervised directly ever achieved much in physics.

Hartle remembers the lectures as being, like all Feynman lectures, brilliant and memorable, giving the students a feel for physics at the cutting edge of research. His own career was particularly influenced by Gell-Mann, John Wheeler and others, and would, he says, probably have followed the same path even without the Feynman lectures on gravity. But the major idea which Feynman introduced into thinking about gravity at around that time was the perturbation technique that had previously been developed in the context of QED. It was another example of Feynman coming up with the right tool for the job from his extensive kit of mathematical techniques.[1]

Two of the other students who attended those classes, Fernando Morinigo and William Wagner, made notes which were edited and reproduced for sale in the Caltech bookstore, where they have been purchased by generations of students ever since. Thirty years later, they were turned into a book by Brian Hatfield.[2] You might think that this is a piece of cynical exploitation of the 'Feynman industry' that has sprung up since he died, like the repackaging of old recordings by dead rock stars. But you would be wrong. Although the work is extremely technical in parts, it is also more than ever of relevance to serious students of gravity; *Feynman Lectures on Gravitation* also contains a strong flavour of Feynman the teacher at work, and some astonishingly prescient insights.

For those serious students, perhaps the most important feature of the book is the way in which Feynman develops the theory of gravity from scratch, using the standard techniques of quantum physics. We saw before how he had found that the entire classical theory of electromagnetism, including Maxwell's equations, could be derived starting out from a quantum description of interactions between particles that have charge, involving the exchange of photons, which are regarded as massless particles with one unit of quantum 'spin'. In the first part of his lecture course, Feynman showed that the entire classical theory of gravity, including Einstein's equations of the General Theory of Relativity, could be derived starting out from a quantum description of interactions between particles that have mass, involving the exchange of gravitons, which are regarded as massless particles with 2 units of quantum spin. The situation is more complicated than in QED because the gravitons can interact with each other, as well as with massive particles, so renormalization doesn't work. The other difference is that with gravity, unlike the case in electromagnetism (where like charges repel and unlike charges attract), like gravitational 'charges' (that is, masses) attract one another. But the philosophical approach is just the same, and provides yet another example of the way in which fundamental truths in physics can usually be described in more than one mathematical formalism.

In an introductory commentary to *Gravitation*, Brian Hatfield emphasizes that this need to develop his own understanding of any problem he worked on was typical of Feynman, who for many years had the slogan, 'What I cannot create, I do not understand' written on the corner of one of the blackboards in his office. If Feynman wanted to study gravity, the only way he could do it was by creating his own theory of gravity, not by looking for ways to improve Einstein's theory. Hatfield describes Feynman's approach to gravity theory as being 'from the bottom up, instead of from the top down', in contrast to the top-down approach of Einstein himself, based on a geometrical description of spacetime in four dimensions, which is the way students are usually introduced to the subject.[3]

Hatfield also comments on Feynman's sometimes cavalier way with conventions such as the way indices are written in mathematical equations: 'Feynman once told me that getting minus signs, and factors of *i*, 2 and pi down correctly was something to be bothered with only when it came time to publish the result.' In the first six of the gravity lectures, Feynman wrote almost every index down (for example, x_i) where the usual convention is to have them up (in this case, x^i). This doesn't matter at all as long as the use is consistent, but Hatfield has restored the more familiar convention in the book, where he mentions the first time he saw Feynman's van, in 1981 in a parking lot at Caltech. This was the famous van covered in Feynman diagrams, and he knew who it belonged to because 'the diagram on the back, the only diagram with labels, had all indices in the down position . . . After looking in one of the windows of the van and seeing a bale of hay in the back, my suspicion that the van was Feynman's was confirmed.' (There was a perfectly logical explanation for the bale of hay, since Michelle was a keen horserider; but to Hatfield only Feynman would be driving around campus with a bale of hay.)

We should stress that there is still no completely satisfactory quantum theory of gravity. The approach pioneered by Feynman works very well at reproducing the successes of Einstein's approach in describing the Universe at large, the orbits of the planets around the Sun and so on. But, also like Einstein's version, it is less successful in providing a description of what goes on in the true quantum realm, at very high energies and over very short distances. Nevertheless, the successes are striking, not least because gravity is so weak. The electrical force between two electrons, for example, is a little over 4×10^{42} times as strong as the gravitational force between the same two electrons. Because of this, you need to put a lot of particles together in one lump before their combined gravitational influence on any one particle in the lump is as strong as the influence of neighbouring particles on each other produced by electromagnetic forces. In the first of his lectures on gravity, in which he provides a broad overview of the subject, this leads Feynman to consider, in a spirit of open-mindedness, some extreme possibilities. 'I would like to suggest', he says, 'that it is possible that quantum mechanics fails at large distances and for large objects. Now, mind you, I do not say that I think quantum mechanics *does* fail at large distances, I only say that it is not inconsistent with what we do know.' And he explains that in this context a 'large' object would be one with a mass of about one hundred-thousandth of a gram, containing about a billion billion particles. In an aside to his main theme, talking in 1962, he says that we must 'not neglect to consider' that it is possible for quantum mechanics to fail on this scale, because of some process involving gravity, and that this could resolve such puzzles as the 'Schrödinger's cat paradox'.

The 'paradox' is actually a kind of *reductio ad absurdum* which Erwin Schrödinger put forward in 1935 to show how ridiculous the standard interpretation of quantum mechanics was (this, remember, was after Schrödinger had said he didn't like quantum mechanics and wished he'd never had anything to do with it). The puzzle concerns an (imaginary!) cat locked up in a room with a quantum device that has a 50:50 chance of triggering a cat-killing mechanism. Because the so-called Copenhagen Interpretation (developed by Niels Bohr and others at the end of the 1920s) says that it is the act of looking to see whether the quantum device has been triggered or not which 'collapses the wavefunction' and makes it decide what state it is in, it can be argued that the cat itself is neither dead nor alive, but exists in a 'superposition of states' until somebody opens the door of the room and takes a look.

Although it looks ridiculous when pushed to such extremes, nevertheless (and in spite of Schrödinger's best efforts) this Copenhagen Interpretation, involving a role for the observer in determining the behaviour of the quantum world just by looking at it, is the standard picture that has been taught since the 1920s. So the idea of a gravitational (or any other) explanation for the distinction between the everyday world and the quantum world, removing the role of the observer, has obvious appeal; it has recently been revived and is widely discussed today (although seldom, if ever, with credit to Feynman's insight).[4]

As for the cat puzzle itself, Feynman clearly expresses his objections to the conventional explanation of how the quantum world works, the Copenhagen Interpretation which holds that it is the act of observation which forces the quantum world to choose one reality out of the array of probabilities described by the wave function. To Feynman,

> This is a horrible viewpoint. Do you seriously entertain the thought that without an observer there is no reality? Which observer? Any observer? Is a fly an observer? Is a star an observer? Was there no reality in the universe before 10^9 BC when life began?[5]

He also deliberates on the 'many worlds' idea, that the Universe constantly splits into slightly different versions of reality every time it is faced with a 'choice' at the quantum level, and points out that according to the conventional understanding of quantum mechanics this is the only way to describe the entire Universe, in terms of 'a complete Monster Wavefunction', because there is no outside observer to 'collapse the wavefunction' and bring one of the possible quantum realities into a unique existence. This is precisely the line that has been taken up by a leading school of cosmologists in recent years, leading to a quantum description of the Universe which is based on a combination of the many worlds idea and the sum over histories approach; one of the leading lights of this school has been James Hartle, one of those students from Feynman's gravity course. What Feynman himself described in 1963 as 'very wild speculations' are part of mainstream discussions today.

Feynman's own discussions of the cosmological and astronomical implications of his work look, with hindsight, even more prescient. He stresses the importance of the fact that everywhere we look in the Universe we see objects that are far from equilibrium, with hot stars pouring energy out into a cold Universe. Studies of non-equilibrium states are also at the forefront of physics today, where researchers are trying to find out how complexity (including life) can arise out of chaos.[6]

But perhaps the most staggering insight into Feynman's feel for physics comes from the way he pre-empted, 20 years in advance, the theory of the origin of the Universe that now goes by the name 'inflation'. The key to this picture of how something appeared out of nothing at all some 15 billion years ago is the realization that the energy in a gravitational field associated with an object of mass m is not only negative, but exactly balances the rest mass energy of the particle, mc^2. The way to picture this is to imagine taking all the constituents of the mass m and spreading them out until they are infinitely far apart. Because the gravitational force between the particles goes as one over the separation squared, when the separation is infinite the force goes as one divided by infinity, which is zero. So the constituents can do no work on each other – they cannot tug each other about – when they are infinitely far apart, which means that the energy of the gravitational field is zero in that situation.

Now, imagine the constituents falling together to make the mass m.* Because gravity is an attractive force, the constituents release energy as they fall together. This is why collapsing clouds of gas in space get hot in the first place, as they shrink down to form

*Of course, it would take for ever for them to fall from infinity; strictly speaking, we should talk about what happens as the separation 'tends to' infinity, but the conclusions from the proper mathematical treatment are the same as in our simplistic example.

protostars; energy comes out of the gravitational field as the cloud collapses, and this heats the cloud up. But if you start with zero energy, and take energy out of the field as the object collapses, that means that for everyday objects the energy in the associated gravitational field is negative! Indeed, if you were to collapse the object all the way down to a mathematical point (a singularity), the energy of the associated gravitational field would indeed be $-mc^2$. Interestingly, although the exact balance between rest mass energy and gravitational energy comes naturally out of the General Theory of Relativity (either Einstein's version or Feynman's version), in Newtonian theory the gravitational field ends up with infinite negative energy, which would be even harder to comprehend.

This curious fact – the balance between mass energy and gravitational energy – had been known (as a mere curiosity) for about 20 years by the time Feynman gave his lectures on gravitation. Back in the 1940s, on a visit to Einstein in Princeton, the pioneering cosmologist George Gamow casually mentioned, while they were out walking, that a colleague, Pascual Jordan, had realized that a star might be made out of nothing, since at the point zero its negative gravitational energy is numerically equal to its positive rest mass energy.

> Einstein stopped in his tracks, and, since we were crossing a street, several cars had to stop to avoid running us down.[7]

In spite of its impact on Einstein, Jordan's idea was regarded as no more than a curiosity, and probably Feynman had never heard of it. Certainly nobody had thought of applying it to the Universe as a whole. In 1962, the idea that the Universe had a definite beginning – the Big Bang – was still very much in doubt, and the famous 'three degrees Kelvin' background radiation which is regarded as the echo of the Big Bang had yet to be discovered. The rival Steady State hypothesis, which holds that the Universe has existed in more or less its present form for ever, was still very much a viable alternative, and was, indeed, discussed at length by Feynman in his gravity lectures. But he was also deeply impressed by the possibility 'that the total energy of the universe is zero'. He pointed out that 'it is exciting to think that it costs *nothing* to create a new particle', and went on to say that:

> We get the exciting result that the total energy of the universe is zero. Why this should be so is one of the great mysteries – and therefore one of the most important questions of physics. After all, what would be the use of studying physics if the mysteries were not the most important things to investigate?[8]

All of this requires that the amount of matter in the Universe should be just enough to match the so-called 'critical' density, for which spacetime is described as being flat and the Universe is just poised on the knife edge between expanding for ever or one day recollapsing into a big crunch. For the critical density (and only the critical density) the Universe does indeed tend to disperse itself to infinity, like our imaginary mass *m*, and end up hovering in a stationary, infinitely spread-out state.

This requires the presence of large amounts of 'dark matter', not yet directly detected but now very fashionable in cosmology, not least because improved observations of the way galaxies move have shown that they are indeed being tugged on by the gravitational influence of large amounts of dark stuff. But this view was distinctly unfashionable in 1962. That didn't worry Feynman, who said that 'the critical density is just about the best density to use in cosmological problems', largely because it is the density for which the creation of matter 'costs nothing'. But he still cautioned against accepting the idea just because it was so attractive:

> It is exciting to speculate that it indeed is the 'true' density – yet we must not fool ourselves into thinking that a beautiful result is more reliable simply because of its 'beauty,' which is in part an artificial result of our assumptions.[9]

The idea that the Universe might have appeared in this way out of nothing at all passed completely unnoticed by the cosmologists, and was reinvented, independently, by Edward Tryon, of City University in New York, in 1973. Nobody took much notice even then (although the idea was published in the journal *Nature*), because it seemed that a tiny seed of the Universe, created out of nothing but containing as much mass as our entire Universe, would immediately collapse back into a singularity because of its own intense gravitational pull. But at the end of the 1970s and in the early 1980s, several researchers (most notably, Alan Guth in the

United States and Andrei Linde in the Soviet Union) developed the idea of inflation, a kind of antigravity that would operate in the first split-second of the existence of the Universe, whooshing it up in size from something much smaller than a proton to something roughly the size of a grapefruit, and giving it so much outward thrust that even after inflation switched off and gravity began its work of pulling things back together the expansion would continue, slowing all the time, for tens of billions of years, allowing a Universe like the one we see around us to develop.[10] None of these pioneers seems to have been aware that a central plank in their platform, the possibility of creating a Universe out of nothing at all because of the balance between mass energy and gravitational energy, had first been suggested by Richard Feynman in 1962. To someone (JG) who studied cosmology in the 1960s, and followed and reported on the developments in the 1970s and 1980s that led to the inflationary scenario becoming accepted as the standard paradigm, it was a breathtaking revelation to open up *Gravitation* in the summer of 1995 and find the extent of the insights provided by Feynman so long ago.

Perhaps this should not have come as quite such a surprise, though, because thanks to Willy Fowler I already knew about one of Feynman's astrophysical insights, which is highlighted in a foreword to *Gravitation*, by John Preskill and Kip Thorne.

As we have already mentioned, it was early in 1963, shortly after the objects now known as quasars were discovered, that Fred Hoyle gave a seminar at Caltech in which he suggested that these objects might be superstars, and both Hoyle and Fowler were astonished when Feynman immediately pointed out that effects described by the General Theory of Relativity would make such supermassive stars unstable. The background to this 'bolt from the blue', as it seemed to Fowler and Hoyle, has now been pieced together by Preskill and Thorne, and part of the story is presented by Feynman in Lecture 14 of the gravity series. It seems that early in January 1963 Feynman visited the astrophysicist Icko Iben, then working at the Kellogg Radiation Laboratory at Caltech, and showed Iben the basic set of equations required to describe the structure of a star, taking full account of the General Theory of Relativity. Feynman had worked these out himself, from first principles. He asked Iben how astrophysicists used the equivalent, much simpler, Newtonian equations to make theoretical models of the behaviour of ordinary stars, where general relativistic effects are not important. Iben showed him. Those classical calculations of stellar structure represented the culmination of about 30 years' work by astrophysicists. A few days later, Feynman came to see Iben again. 'Feynman flabbergasted me,' Iben recalls, 'by coming in and telling me that he had [already] solved the . . . equations. He told me that he was doing some consulting for a computer firm and solved the equations in real time on what must have been that generation's version of a workstation.'[11] A couple of days after that, on 28 January, Feynman delivered Lecture 14 in the gravity series. It describes a fully general relativistic model of supermassive stars, still valid today, although Feynman's interpretation of his calculations is not quite correct. It was a few weeks after this lecture that Hoyle gave the now-famous talk at Caltech.

Impressive though all this is, it was to some extent peripheral to the main object of Feynman's investigation of gravity, which was to develop a complete quantum theory. Without ever completing the work, he pointed the way very clearly for the next generations of researchers. Just as in QED, in quantum gravity Feynman diagrams without any 'loops' describe interactions that follow the rules of the classical theory. In QED, you can add one loop to the diagram, and calculate the resulting quantum correction, then add two loops, then three, and so on, developing an ever more accurate calculation (for example, of the magnetic moment of the electron) as long as you have sufficient patience and enough computer power to carry out the calculations (this is the perturbation approach mentioned by Hartle). For gravity, largely because of the way in which gravitons can interact with each other, even setting up the right equations to solve is more difficult, and when you do set them up they are plagued by infinities. Feynman only ever got as far as making the one-loop correction – itself a considerable achievement, accomplished in the summer of 1962 (it was probably this success which encouraged him to give the lectures on gravitation a few months later). Importantly, Feynman found that in order for this approach to work at all, he had to include the influence of 'ghost' fields, responsible for the presence of particles which exist *only* as self-contained loops in the Feynman diagrams and have no 'real' existence at all. It was this insight which enabled others to take the approach further, developing the techniques to describe how effects involving larger numbers of loops ('higher-order' calculations) should be included in the calculations, using path integral techniques. According to University of Texas researcher Bryce DeWitt, one of the leading investigators of quantum gravity today, 'his work on quantum gravity ultimately had great impact on the standard model and on the quantization of gauge fields in general . . . people are well aware of his contribution'.[12] Modern quantum gravity theory is one of the most exciting developments in theoretical physics, and it has Feynman's fingerprints all over it. Feynman himself was happy with his achievement:

> I feel I have solved the [problem of the] quantum theory of gravity in the sense that I figured out how to get the quantum principles into gravity. The result is a nonrenormalizable theory, showing it to be an incomplete theory in the sense that you cannot compute anything. But I am not dissatisfied with my attempt to put gravity and quantum mechanics together. I accept whatever consequences that this putting together produces, mainly that it can't be renormalized. I was slightly disappointed that I did it only to lowest order. I could not figure out what to do with arbitrary numbers of loops, which was later solved by others, but I was not dissatisfied with that. The fact that the theory has infinities never bothered me quite so much as it bothers others, because I always thought that it just meant that we've gone too far: that when we go to very short distances the world is very different; geometry, or whatever it is, is different, and it's all very subtle.[13]

This wasn't just an afterthought by Feynman. In *Gravitation*, he had already said that the Lagrangian that emerges from Einstein's General Theory of Relativity is merely an 'effective Lagrangian' that describes the low-energy behaviour of a more fundamental theory, in much the same way that Newtonian gravity is an effective theory that describes the behaviour of objects under even less extreme conditions where we do not even need to take account of general relativistic effects. The more fundamental theory that underpins both the General Theory and Newtonian gravity would operate, he suggested, on the tiniest scale, the so-called 'Planck length'.

The Planck length is a number, with the dimensions of length, that can be derived from the three fundamental constants of physics (the constant of gravity, the speed of light and Planck's constant), which respectively relate to gravitation, electromagnetism, and the quantum world. There is only one length that can be derived from a combination of these numbers, and it has a value of about 10^{-33} centimetres. The Planck length is the length scale on which gravity, electromagnetism and quantum phenomena are on an equal footing, and it is in a sense the smallest possible distance that can exist, the 'quantum of length'.

One of the most dramatic developments in theoretical physics came in the mid-1980s, when, almost by accident, theorists found a theory which describes what is going on at these astonishingly small length scales, and which automatically gives rise to gravity as we know it, in just the way that Feynman predicted. It is known as superstring theory, and is still the best all-embracing theory of the origins of particles and gravity that we have.

The central idea of all string theories is that the fundamental entities of the physical world are not point-like objects, the way we are used to thinking of leptons and quarks, but have some extension in one dimension, like a line drawn on a piece of paper. The extension is very small, comparable to the Planck length, but it is definitely not zero. Even so, there is no prospect of ever being able to detect one of these strings – it would take 100 billion billion of them, laid end to end, to stretch across the diameter of a proton. This means that the size of such a string, compared to the nucleus of an atom, is equivalent to the size of a nucleus, compared to the size of the Sun. In the 1970s, some mathematicians dabbled with calculations describing the behaviour of such strings, but this was more because they were interested in the mathematics involved for its own sake, rather than through any suspicion that the equations they were playing with might describe the real world. Then, in the 1980s, they began playing with an improved version of the idea, called superstring theory, and came up with results that made the physicists begin to sit up and take notice.

In superstring theory, the fundamental entities are thought of as little lengths or loops of vibrating string, with the kinds of properties we associate with 'fundamental particles' (such as the charge on the electron) either tied to the ends of the open strings or associated with the way the strings vibrate. A closed loop of string – like a tiny, vibrating elastic band – is fundamentally different from an open string, but any theory that describes open strings automatically includes closed loops as well. To their surprise, when mathematical physicists calculated the properties of these closed loops of string in the mid-1980s, they found that they were equivalent to massless particles with 2 units of quantum spin. In other words, gravitons. Superstring theory *predicts* the existence of gravitons, and Feynman had shown, 20 years before, that gravitons are all you need to produce a theory of gravity identical, on the appropriate energy and distance scale, to the General Theory of Relativity.

It gets better. The infinities that plagued earlier attempts to develop a quantum theory of gravity do not arise in superstring theory, which is both mathematically self-consistent and finite. It has all the characteristics of the new theory required to describe what goes on at very short distance scales that Feynman alluded to.

Feynman's fascination with Mach's Principle also came into the lectures on gravitation, and also provides a direct link with current developments in physics. The idea that the inertia of an object arises as a result of gravitational interactions with very distant objects clearly has a family resemblance to Feynman's old idea that the radiation resistance experienced by a charged particle – a kind of electrical inertia – arises as a result of electromagnetic interactions with very distant charged particles. In his gravity lectures, Feynman stops short of invoking a role for advanced gravitational interactions to account for inertia in the way that he and Wheeler had once invoked a role for advanced electromagnetic interactions to describe the forces acting between charged particles. Rather, he concludes his discussion of Mach's Principle with another memorable piece of his philosophy of science:

> The answer to all these questions may not be simple. I know there are some scientists who go about preaching that Nature always takes on the simplest solutions. Yet the simplest solution by far would be nothing, that there should be nothing at all in the universe. Nature is far more inventive than that, so I refuse to go along thinking it always has to be simple.[14]

It is easy to imagine, in a speculative way, a kind of 'explanation' of Mach's Principle involving advanced and retarded gravitational interactions criss-crossing the Universe in the way that electromagnetic interactions move forwards and backwards in time in the Wheeler–Feynman theory of radiation. But it was only in 1993 that this kind of approach was put on a secure footing, by the work of Shu-Yuan Chu, of the University of California. Chu has developed a model of how to do quantum mechanics in the presence of gravity, which combines some of the latest ideas in particle physics (including superstrings) with a time-symmetric Wheeler–Feynman description of gravity and inertia.

Following Feynman's example, Chu does away with the concept of a 'field', and works entirely in terms of particles (photons, gravitons and the like) being exchanged between other particles in a time-symmetric way. He suggests that this continuous feedback, on the smallest scale, builds up what we think of as continuous fields (such as gravity) as the average over all the interactions involving little pieces of matter. The averaging takes place on a scale that is large compared with the size of a string – but that still means that it happens on a scale far smaller than the size of a proton, so that our instruments are quite incapable of detecting it directly, and we only perceive a smooth field. Chu says that the effect would be like admiring a superbly woven tapestry from across the room, where it seems to make a smoothly continuous picture; only when you look at it up close could you see the individual threads that go together to make up the tapestry. And the kind of averaging you have to do, to get the smooth picture that we are familiar with, is, of course, the averaging involved in Feynman's path integral approach to quantum physics.

This approach explains the origin of inertia, and Mach's Principle, in the context of superstring theory, using exactly the mathematical formalism of Wheeler–Feynman electrodynamics. It also implicitly includes the Wheeler–Feynman theory of electrodynamics and the origin of radiation resistance. It's a rather nice bonus that there is now good evidence that, as Feynman suspected, the Universe does contain the critical density of matter, making it flat, which ensures that there is enough matter in the future to provide the 'echoes' needed for the advanced and retarded waves to match up in the required way, without having to introduce any extra bells and whistles into the theory. Chu confessed to feeling more than a little nervous at going public with such an outrageous idea, that advanced interactions ('messages from the future') might play a fundamental part in determining the structure of the world as we perceive it.[15] But what he didn't know at the time he developed his model was that this outrageous idea had already been revived, in the context of 'ordinary' quantum mechanics (without strings) back in 1986, by John Cramer, of the University of Washington, Seattle.[16]

Cramer picked up on a rather peculiar feature of the Schrödinger equation itself, a feature which has long been known about and largely been ignored. Way back in 1927, in the early days of quantum mechanics, the pioneering astrophysicist Arthur Eddington pointed out that the quantum probabilities which are so important in making calculations of the behaviour of entities in the quantum world are 'obtained by introducing two symmetrical systems of waves travelling in opposite directions of time'.[17] The situation is very similar to the way in which there are two sets of solutions to Maxwell's equations of electromagnetism, but with an important difference. With Maxwell's equations, you can carry out the calculations either by using just one set of solutions, and completely ignoring the other set of solutions; or you can choose to use a mixture of half advanced and half retarded waves. With Schrödinger's equation, you have no choice. You *always* have to use a mixture of advanced and retarded waves to calculate the probabilities.

It happens like this. Schrödinger's wave equation involves what mathematicians refer to as complex numbers, in which the square root of -1, denoted by i, appears. In spite of their name, there is nothing particularly difficult about handling complex numbers; as we mentioned earlier, the name actually indicates that they are made up of two components, typically with a form like $(x + it)$, instead of 'simply' being made up of either ordinary numbers (like x) or so-called 'imaginary' numbers (like it). And the need for two components to describe these numbers can be pictured in terms of the little arrows described in Chapter 4. As in this example, the 'imaginary part' of a complex equation describing the behaviour of a wave is linked to the time, denoted by t. The whole thing describes what is known as the amplitude for a particular interaction, or, say, for one route which might be taken by an electron through one hole in the experiment with two holes. But remember that in order to calculate the *probability* of a particular quantum event, you have to take the *square* of the amplitude; and this is where things get interesting.

Everyone knows how to make the square of an ordinary number, like x. You simply multiply it by itself, $x \times x$. But this is not the way you make the square of a complex number, like $(x + it)$. Instead, you multiply it by something called its complex conjugate, in which you change the sign in front of the imaginary part of the number, making it $(x - it)$, so that for the square you get $(x + it) \times (x - it)$. Schrödinger's equation is just a little more complicated than this simple example, but the principle is the same. And by reversing the sign in front of the t in Schrödinger's equation, you have automatically selected the opposite version of the equation, describing a wave moving backwards in time. Extending the analogy used earlier, the rotating arrow that defines the phase of the wave is rotating in the opposite direction. Every time any physicist has ever calculated a quantum probability using Schrödinger's equation in this way, they have automatically been taking account of both the advanced and the retarded waves in their calculation.

So, as Cramer pointed out in 1986, the quantum world can be described along exactly the lines of the Wheeler–Feynman theory of radiation, in which the advanced and retarded waves combine to produce an effective 'action at a distance' which takes no time at all. The way to picture this is to imagine standing outside of time, and watching what goes on as if it were happening sequentially, but remembering that it is really all happening at once. On this picture, a particle which has the potential to get involved in a quantum interaction sends out what Cramer calls an 'offer' wave, moving symmetrically in both directions of time, into the past and into the future. There is no distinction between the roles of past and future in this picture, but for our peace of mind just concentrate on the wave going out, in all directions, into the future. Out in the Universe at large, the wave triggers a response – indeed, it may trigger many responses, from many other particles. In each case, the triggered particle sends out a 'confirmation' wave, also into the past and into the future, indicating its ability to take part in the interaction. All of the confirmation waves travelling back in time arrive at the originating particle at the same instant that it made the original offer, and it 'chooses' one of the confirmation waves, in accordance with the familiar rules of quantum probability, to take part in the transaction. Everywhere else, all of the waves cancel each other out, leaving a completed transaction between two particles (see Figure 16), made up from both solutions to Schrödinger's equation and forming a firm handshake across spacetime. From the 'point of view' of the waves themselves, the whole thing takes zero time.

The classic example of the experiment with two holes makes

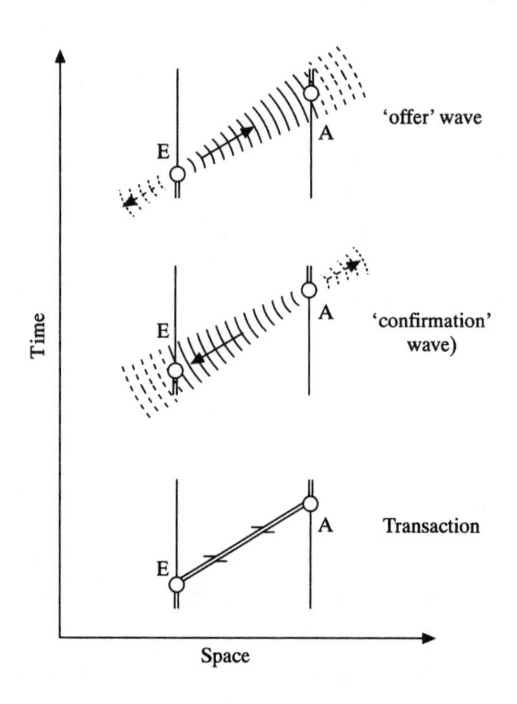

Figure 16. John Cramer has developed the idea of waves moving forwards and backwards in time (see Figure 5) to describe quantum interactions. Because the offer wave travels forwards in time and the confirmation wave travels backwards in time, the transaction takes no time at all to complete. This provides an explanation of quantum mysteries such as how electrons 'know in advance' whether one or both holes are open in the experiment with two holes (see Figure 4). E = emitter; A = absorber.

the situation clear. In this case, the offer wave goes out through both holes, before the electron ever sets out on its journey. The confirmation wave also comes back through both holes – indeed, confirmation waves come back by any possible route from anywhere the electron could possibly go, just like the crazy way in which light bouncing off a mirror behaves in QED. Just one confirmation wave is accepted by the electron, so the electron itself actually goes by one route to its destination on the detector screen. But its position on the detector screen – the point where it makes a blob of light – is determined by the whole structure of the experiment, taking account of both holes to create an interference pattern as more and more electrons are faced with the equivalent choices. Crucially, if one of the holes is covered up, then this theory predicts that the behaviour of the electrons and the pattern they make on the detector screen will change in exactly the way it is seen to change in experiments. As well as making full use of both advanced and retarded waves, nature really is, on this picture, carrying out a 'sum over histories' to determine where the electron ends up.

This view resolves the famous 'Schrödinger's cat paradox'. At the *beginning* of the experiment, advanced waves from the future present the quantum system with a 'choice' of a dead cat or a live cat, and the transaction confirming which choice will become real is made, on a 50:50 basis, before anything happens. The fate of the cat is indeed sealed by quantum probabilities, but it is sealed from the outset, with no need for a 'superposition of states', and no need for the role of an observer in creating reality in the way that Feynman ridiculed. And *all* of the puzzles and mysteries of the quantum world dissolve away into transparency in this way, because any quantum entity involved in any quantum experiment really does 'know' about the entire structure of the experiment, and the entity's ultimate fate, before anything happens at all in a human timeframe. As Cramer puts it:

> If there is one particular link in [the] event chain that is special, it is not the one that ends the chain. It is the link at the beginning of the chain when the emitter, having received various confirmation waves from its offer wave, reinforces one of them in such a way that it brings that particular confirmation wave into reality as a completed transaction. The atemporal transaction does not have a 'when' at the end.[18]

Cramer called this the 'transactional interpretation' of quantum mechanics. In a sense, it is 'only' an interpretation – this way of looking at things makes no predictions about the behaviour of the quantum world that differ from the predictions made by the Copenhagen Interpretation, or by Feynman's path integral formalism. But this, of course, is a strength of Cramer's interpretation, because that means that his picture, like the others, agrees with all the thousands of experimental results concerning the quantum world that have been obtained in the past 70 years and more. The great thing about the transactional interpretation is that it provides you with an easy way to get a picture of what is going on in the quantum world, without mysteries such as cats that are dead and alive at the same time, or electrons that go through two holes at once, at the cost of accepting the reality of the advanced waves. But since physicists have been implicitly accepting the reality of advanced waves every time they have used Schrödinger's equation to calculate quantum probabilities since 1926 (and some of them, like Eddington, even knew what they were doing), that seems a pretty small price to pay!

This is an example of the direct influence of Richard Feynman on modern physics, with researchers picking up on his ideas and developing them in new ways – in this case, half a century after he first got involved with describing the behaviour of the world with the aid of advanced waves. It is a good note on which to end our discussion of modern science, because it combines one of Feynman's earliest pieces of scientific research with one of the latest ideas in thinking about the quantum world, to resolve what he himself called the central mystery of the experiment with two holes; and, through the work of Chu, it provides a possible explanation of the physics behind one of the mysteries of the Universe, Mach's Principle, that both puzzled and intrigued Feynman for decades.

But there is another way in which Feynman continues to influence modern physics, and that is through his approach to physics – indeed, to life in general – epitomized by his teaching. Feynman himself suggested that his most important contribution, in the long term, would turn out to be the *Lectures*, which describe his approach to physics.[19] With the evolution of science, it is impossible to predict how long Feynman's scientific contributions will last, and in what form. But by teaching people how to think, insisting on scrupulous honesty and integrity, never fooling yourself and always rejecting any theory, no matter how cherished, if it disagrees with experiment, and, above all, inspiring an awe and appreciation for nature and a love of science, Feynman made a mark on science which will last whatever happens to the science itself as new experiments to test its predictions are carried out. David Goodstein says that:

> His scientific contributions were profound, they are not ordinary. They are not similar to other people's. He imposed his personality and his view on the world of science; he reformulated quantum mechanics, he virtually reinvented it. And he gave it to us in a form that's still widely used throughout theoretical physics, in every field.[20]

As Laurie Brown and John Rigden put it in the introduction to *Most of the Good Stuff*, 'there is an important sense in which all modern physicists are Feynman's students'. And they all miss him.

Notes

1. James Hartle, telephone conversation with JG, December 1995.
2. *Feynman Lectures on Gravitation*, by Richard Feynman, Fernando Morinigo & William Wagner, edited by Brian Hatfield, Addison Wesley, 1995. Hereafter referred to as *Gravitation.*
3. For a description of Einstein's work and the traditional approach, see *Einstein: A Life in Science* (Michael White & John Gribbin, Plume, New York, and Simon & Schuster, London, 1995).
4. See *Schrödinger's Kittens.*
5. *Gravitation.*
6. See, for example, *At Home in the Universe* (Stuart Kauffman, Oxford University Press, New York, 1995), and Kauffman's earlier book *The Origins of Order* (Oxford University Press, New York, 1993).
7. George Gamow, *My World Line*, Viking, New York, 1970.
8. *Gravitation.*
9. *Gravitation.*
10. There are sound observational reasons why inflation is taken seriously, which we do not have space to go into here. See *In the Beginning* (John Gribbin, Penguin, London, and Little, Brown, New York, 1993).
11. *Gravitation*, Foreword.

12. Comment to JG, December 1995.
13. Mehra.
14. *Gravitation.*
15. Letter to JG, March 1994.
16. See *Schrödinger's Kittens.*
17. Arthur Eddington, *The Nature of the Physical World*, first delivered as a series of lectures in Edinburgh in 1927, published by Cambridge University Press in 1928 and in many editions down the years, including as an Ann Arbor Paperback by University of Michigan Press, 1958.
18. John Cramer, 'The Transactional Interpretation of Quantum Mechanics', *Reviews of Modern Physics*, volume 58, p. 647 *et seq.*, 1986. If you have Internet access and a web browser, you can find this paper at http://mist.npl.washington.edu/npl/int_rep/tiqm/TI_toc.html.
19. David Goodstein, interview with JG, April 1995.
20. See note 19.

Epilogue: In search of Feynman's van

Seven years after Richard Feynman died, one of us (JG) visited Caltech for the first time. One reason for the visit was to give a talk about the transactional interpretation of quantum mechanics, outlined in Chapter 14, which draws so strongly on Feynman's own unusual ideas about the nature of electromagnetic radiation, now more than half a century old. It was, to say the least, an unusual feeling to be talking not just from the spot where Feynman himself used to lecture, but about his own work. And when, during the question period at the end of the talk, the discussion moved on to QED, the dreamlike quality of the occasion intensified – an audience at Caltech, of all places, was asking *me* to explain QED to them!

But the main purpose of the visit was to fill in the background to the Feynman legend in preparation for writing this book, visiting the places where he used to work and meeting the people he used to work with. In the spring of 1995, after an unusually wet late winter, the Caltech campus seemed to be the ideal place for a scientist (or anyone else) to work. With temperatures in the eighties and a cloudless sky, the green open spaces of the campus, shaded by trees and decked with colourful flowerbeds, offered a calm environment highly conducive to gentle contemplation about the mysteries of the Universe. I was reminded of a visit to Laugharne, in South Wales, to the modest building where Dylan Thomas used to work, looking out over the spectacular views and thinking, 'if I'd lived here, even I might have become a poet'; I may not be much of a physicist, but the atmosphere at Caltech makes you think, 'if I worked here, even I might have one or two good ideas'. And then you think about the people who have worked there, including

Feynman himself, Murray Gell-Mann, whose room was separated from Feynman's only by Helen Tuck's office, and Kip Thorne, one of the two or three leading experts on the General Theory of Relativity, still working at Caltech but not too busy to take time off to discuss black holes, time travel and Feynman. And then you think, 'well, maybe my ideas wouldn't be *that* good'.

The point about Caltech, in academic terms, is that not only does it bring out the best work from its scientists, it also (partly for that reason) attracts the best scientists. So what you end up with is the best of the best. There are always top people eager to become part of the Caltech scene; but Feynman himself has never been directly replaced, even though, after his death, a committee was set up to seek a replacement. They failed to find one, because there is nobody like Feynman around today – just as there never was anybody like Feynman, except Feynman himself, around before.

There is no formal memorial to Feynman. No grand building, or statue. Even his grave, shared with Gweneth in Mountain View Cemetery in Altadena, is very simple. His real memorial is his work, his books and the video tapes on which he can still be seen, lecturing in his inimitable style, making difficult concepts seem simple. But there is one artefact which strikes a curious resonance with anybody who has ever heard of Feynman, and which I had been urged, by a friend who knows next to nothing about science but still regards Feynman as a hero for our time, to track down while I was in Pasadena.

The opportunity came at the end of a long talk with Ralph Leighton, in the lobby of my hotel on Los Robles Boulevard. My host in Pasadena, Michael Shermer of the Skeptics Society, sat in with us for a conversation which ranged not only over Feynman's life and work, but also over the reaction of the world at large to his death, and the reaction of Feynman's family and friends to the way he had been presented in various books and articles since then. That conversation brought me as close as I could ever hope to get to the man himself, confirming and strengthening the impressions I already had about what kind of person he was, and shaping the book which you now hold. Richard Feynman was indeed, as well as being a scientific genius, a good man who spread love and affection among his family, friends and acquaintances. In spite of the dark period in his life after the death of Arline, he was a sunny character who made people feel good, a genuinely fun-loving, kind and generous man, as well as being the greatest physicist of his generation. And it is that spirit, rather than the physics, which makes people so curious about the artefact – Feynman's famous van, replete with diagrams.

Our conversation with Leighton had been so intense that I hesitated to bring up the relatively trivial question I had promised to ask. But as we walked him back to his car in the spring sunshine, I reminded myself that a promise is a promise. 'By the way,' I said, 'whatever happened to Feynman's van?' 'It's still in the family, so to speak', he replied. Michael Shermer's ears visibly pricked up at the news: 'Where?' 'It needs some work. It's parked out at the back of a repair shop in . . .' and he gave us the name of another part of the Los Angeles urban sprawl, out to the east of Pasadena.

That, I thought, was the end of it. I had no transport of my own in Pasadena, and although I'd kept my promise to ask after the van, I wouldn't be able, as I'd hoped, to get a picture of it for my friend. I had a radio talk show engagement ahead of me, and an early flight out the next morning. But Shermer had other ideas. He offered to drive me over to find the van as soon as I'd finished at KPCC-FM, and seemed at least as eager as I was to make the pilgrimage. A couple of hours later, we were cruising around the location that Leighton had pointed us towards, stopping to call him on Shermer's car phone for directions each time we got lost. Just as the sun was setting, we found the repair shop, parked and walked around the back. There it was. Feynman's van, nose up against the wall, looking slightly battered but still with its decorative paintwork of Feynman diagrams. It had clearly been there for some time, and delicate spring flowers were growing up around its wheels.

We took our pictures and left, congratulating ourselves on completing the 'Feynman tour' successfully. Twelve hours later, I was in San Francisco, and it was only on my return home that I heard from Shermer about the sequel to the story. The next day, he had happily recounted the tale of our search for Feynman's van to a friend who works at the Jet Propulsion Laboratory, a space research centre in Pasadena. The friend, a sober scientist himself, and hardly an obvious science 'groupie', eagerly asked for directions to the repair shop, and went out there the same day, armed with his own camera. Shermer's joke about the Feynman tour has now almost become reality, with a succession of visitors to the relic – and out of all the pictures I brought back from my California trip, the ones that continue to rouse the most interest are the ones of a beaten up old van parked at the back of a repair shop somewhere east of Pasadena.

I'm not sure why, even though I share something of this enthusiasm. But it's nice to know that something which demonstrates so clearly Feynman's sense of fun and irreverence, as well as referring to his Nobel Prize-winning work, still exists. Leighton suggests that the symbol is particularly appropriate, because the van itself is a symbol of Feynman's free spirit, a vehicle of exploration and discovery of the everyday world, while the diagrams symbolize his exploration and enjoyment of the world of physics. Together, they represent what Feynman was all about – the joy of discovery, and the pleasure of finding things out. Leighton says he will make sure the van stays in the family of Feynman's friends, and suggests that it might one day form the centrepiece of a travelling Feynman exhibit. Now that sounds like the kind of memorial even Feynman might have approved of.

Bibliography

FURTHER READING

Books marked with an asterisk include equations and are more technical (at least in parts) than the present book. The rest are accessible at about the same level as this book.

Mainly about Feynman

Richard Feynman & Ralph Leighton, *Surely You're Joking, Mr. Feynman!* (W. W. Norton, New York, 1985).

Richard Feynman & Ralph Leighton, *What Do You Care What Other People Think?* (W. W. Norton, New York, 1988).

Ralph Leighton, *Tuva or Bust!* (W. W. Norton, New York, 1991).

Christopher Sykes (editor), *No Ordinary Genius: The Illustrated Richard Feynman* (W. W. Norton, New York, 1994). This is the book which Feynman's family and friends recommend as providing the most 'true to life' image of the man.

About Feynman's life and work

Laurie Brown & John Rigden (editors), *Most of the Good Stuff* (American Institute of Physics, New York, 1993).

James Gleick, *Genius: Richard Feynman and Modern Physics* (Pantheon, New York, 1992).

*Jagdish Mehra, *The Beat of a Different Drum* (Clarendon Press, Oxford, 1994).

See also the chapters relating to his relationship with Feynman in

Freeman Dyson's books *Disturbing the Universe* (Basic Books, New York, 1979) and *From Eros to Gaia* (Pantheon, New York, 1992).

The impact of Feynman's ideas about tiny machines is discussed in *NANO!*, by Ed Regis (Bantam Press, London, 1995).

About quantum physics, including Feynman's contributions

Richard Feynman, *The Character of Physical Law* (MIT Press, Cambridge, Mass., 1965).

Richard Feynman, *QED: The Strange Theory of Light and Matter* (Princeton University Press, Princeton, 1985).

Richard Feynman, *Six Easy Pieces* (Addison-Wesley, Redding, Mass., 1995).

Richard Feynman & Steven Weinberg, *Elementary Particles and the Laws of Physics* (Cambridge University Press, Cambridge, 1987).

*Richard Feynman, Robert Leighton & Matthew Sands, *The Feynman Lectures on Physics* (Three volumes, Addison-Wesley, Redding, Mass., 1963).

*Richard Feynman, Fernando Morinigo & William Wagner (edited by Brian Hatfield), *Feynman Lectures on Gravitation* (Addison-Wesley, Redding, Mass., 1995).

John Gribbin, *In Search of Schrödinger's Cat* (Bantam, New York, 1984).

John Gribbin, *Schrödinger's Kittens* (Little, Brown, New York, 1995).

*Silvan S. Schweber, *QED and the Men Who Made It* (Princeton University Press, Princeton, 1994).

Also of interest

Alice Kimball Smith & Charles Weiner (editors), *Robert Oppenheimer: Letters and Recollections* (Harvard University Press, 1980).

Michelle Feynman, *The Art of Richard P. Feynman* (Gordon & Breach, Basel, 1995). Many of Feynman's drawings, and a few paintings, chosen and photographed by his daughter, Michelle.

Richard Feynman (edited by Anthony Hey & Robin Allen), *Feynman Lectures on Computation* (Addison-Wesley, Redding, Mass, 1996).

David Goodstein & Judith Goodstein, *Feynman's Lost Lecture* (Jonathan Cape, London, 1996).

FURTHER LISTENING

A delightful and entertaining recording of Richard Feynman drumming with Ralph Leighton and recounting one of his most famous anecdotes, the 'Safecracker Suite', can be obtained from Ralph Leighton at PO Box 70021, Pasadena, CA 91117, USA. The one-hour recording costs £8 on cassette and £12 on CD; proceeds from the sales, after expenses and taxes, benefit cancer research.

Six Easy Pieces (see above) can also be obtained from bookstores with cassettes or CD of the original Feynman lectures on which the books are based.

ELECTRONIC CONNECTIONS

If you have a World Wide Web browser, a search using 'Tuva' or 'Feynman' will throw up many interesting links.

Index

Numbers in italics refer to Figures.

A (axial) interactions 165, 166, 167
absolute zero 155
absorber theory of radiation
  see Wheeler–Feynman theory
ace model 193, 194
action at a distance 74, 76, 81, 82
Albert Einstein Award 148
algebra 13–15
Alix G. Mautner Memorial Lectures 246
American Institute of Physics 256
American Museum of Natural History 2
American Physical Society 109, 117, 118, 163, 169–70
amplitudes 86, 123, 125, 274, 275
Anderson, Carl 42
antimatter 42
antineutrino 133
antiparticles, and particles 127, 133
Aristophanes 180
arithmetic 10–11
Armstrong, Neil 228
Asimov, Isaac 20
astronomy 9, 53, 162
asymptotic freedom 201
atom
  Bohr's model 34–5, 112
  electron cloud 29, 40, 72, 120
  orbits 34, 35
  structure of 29, 34
atomic hypothesis 11
atomic theory 49
'Atoms for Peace' conference (Geneva, 1958) 149
aurora borealis 9, 10, 212
Bacher, Robert 141, 166, 167, 174
Bader, Abram 16, 17, 18, 52, 82
Bardeen, John 63, 158, 159, 260–61
Barnes, Dr Edwin 11
baryons 132, 191, 192, 193, 194, 200
BBC (British Broadcasting Corporation) 178, 256
BCS theory of superconductivity 158, 159
Bell Laboratories 63, 64
Bennett, Charles 223
Bernoulli's equation 50
Bessel functions 113
beta decay 30, 159, 165–6, 190
Bethe, Hans 102, 108, 110, 212
  at Cornell xv, 100–101, 104, 108, 112, 141
  F. impressed with his skills 100–101
  and General Electric 106
  and how nuclear fusion reactions keep the Sun hot 106
  and infinities in QED 106, 107
  at Los Alamos xv, 94, 95, 96, 100, 141
  Nobel Prize 106
Bevatron 193
Big Bang 266
binary pulsar 121
biology 59, 169, 171, 185
Bjorken, James 197, 198, 199, 202
black body 31, 32, 33, 34
black body curve 31, 32, 33
black holes xv, 58, 93, 282
Block, Martin 160–61
Bohr, Niels 34, 58, 95–6, 146, 263–4
Bohr's model of the atom 34–5, 112

Bose, Satyendra 33, 156
Bose–Einstein condensates 157
Bose–Einstein statistics 189
bosons 156–7
  intermediate vector 132, 134, 135
Brattain, Walter 159
Brewster's Angle 144
British Library 171
Broglie, Louis Victor, Prince de 35–6
Brown, Laurie 279
Brown University 207
Byers, Nina 207

calculus 15, 16, 147
Calculus Made Easy (Thompson) 15
Calculus for the Practical Man (Thompson) 15
California Institute of Technology (Caltech) xiv, xv, 281–2
  Athenaeum Club 142, 152, 180, 227
  F. appointed Richard Chace Tolman Professor of Theoretical Physics 173
  F. at xiv, 141, 142, 143, 147, 148, 149, 152–3, 155, 163, 166, 169, 170, 171, 173–5, 179–80, 190, 196, 204, 205, 208, 210, 249, 251, 254, 260, 281, 282
  and F.'s blood transfusions 243, 244
  F.'s relationship with his students xv, 211–16
  Gell-Mann at 160, 165, 190, 193, 195
  and gravitational radiation 164
  Hoyle at 163, 268, 269
  Kellogg Radiation Laboratory 144, 268
  physics teaching 173–8
  Physics X course 211, 228
  stages *Guys and Dolls* 209–10
  stages *South Pacific* 243–4
  Zweig at 193
Carroll, Lewis 111
Case, Ken 163
Catalina island, near Los Angeles 254
Cedarhurst 7, 10, 11, 12
Centre for Research in Physics, Rio de Janeiro 140–41, 143, 146
centrifugal force 9
CERN (European Laboratory for Particle Physics) 135, 181, 193, 201
Chadwick, James 42
Challenger inquiry *see* Rogers Commission
chaos 265
Chrysler 48, 56
Chu, Shu-Yuan 273, 274, 278
City University, New York 267
Cohen, Art 49, 50
Cohen, Michael 213–14
collapse of the wave function 39–40, 264
collapsing stars 164
colour charge 135
Columbia University 16, 22, 105, 160, 161, 165
complex numbers 86–7
complexity 265
computation 223, 226
Connection Machine 226–7
Cooper, Leon 158
Copenhagen Interpretation 263–4, 278
Cornell University, Ithaca, New York xv, 110, 210
  Bethe at *see under* Bethe
  Dyson at 112, 113, 114, 116, 211
  F. asks for his old job back 147
  F. at 100, 102, 103, 104, 137, 138, 139, 141, 142, 143, 155, 178–9, 208, 211, 213
  Messenger Lectures 178–9
cosmic rays 42, 70–71, 160
cosmology 268
Cowan, Glen 250–53, 258
Cramer, John 274, 275, *276*, 277–8
Crick, Francis 185
critical density 267, 273
critical mass 97, 98
Crossman, Bill 49, 50
current algebra 198, 199

Darwin, Charles xv, 227
Davies, Richard 204, 205
Deborah Hospital, Browns Mill, New Jersey 67

Delbrück, Max 169
Deutsche Elektron Syncheoton Anstalt (DESY) 197
DeWitt, Bryce 270
Dickens, Charles 170
diffraction grating 125
Dirac, Paul 36, 101, 137, 186
  first comprehensive textbook on quantum mechanics 43–4, 50, 60, 73
  as F.'s hero 138
  impressed by F. 109
  paper on 'The Lagrangian in quantum mechanics' 83–5
  proves versions of quantum theory to be equivalent to one another 36–7, 88
  and renormalization 136
Dirac equation of the electron 40–42, 77, 93, 105, 106, 109, 120, 159, 166, 186
Disturbing the Universe (Dyson) 112
DNA 147, 169, 185
Dombey, Norman 190
Double Helix, The (Watson) 185
Dresden Codex 205–6, 207
Dyson, Freeman xiv, 178, 242
  and the American Physical Society 118
  at Cornell 112, 113, 114, 116, 211
  and Feynman diagrams 127
  and F.'s O ring experiment 238
  his paper makes quantum electrodynamics accessible 115, 116
  and the Institute for Advanced Study 111, 113
  at the Oldstone Conference 118
  trip to Albuquerque with F. 114–15, 149, 247–8
  works for Bomber Command 112
Dyson graphs 116

Eddington, Arthur 274, 278
Edgar, Robert 169
eightfold way 191, 192, 193, 195
Einstein, Albert xv, 27, 39, 114, 138, 157, 177, 261, 263, 266
  and De Broglie's theory 35–6
  flees from Hitler's Germany 63
  and inertia 75
  last important work 189
  mass-energy equation 28, 74, 190
  theories of relativity *see* relativity
  told of F.'s path integral approach 88–9
  and visualizations 113
  and the Wheeler–Feynman theory 81
elastic theory 219
electricity 25, 143, 173
electrodynamics 80, 136
  time-symmetric 81
electromagnetic waves 30–31, 33
  and Maxwell's equations 25–6
  out of phase *30*
  in phase *30*
electromagnetism 30, 134, 189, 261, 262, 271
  and QED 132
electron(s)
  bouncing off neutrons 117
  effects of spin 104
  equation of the 166
  and experiment with two holes 37, 38, *38*
  going forwards and backwards in time 80, 108, 110
  identified 29, 30
  'jumping' from one orbit in an atom to another 35, 36–7
  magnetic moment of 105, 109, 120–21, 128–9, *128*, 130, 144
  negative-energy 41–2, 43, 77
  negatively charged 29, 30, 41, 42, 71
  as particles 36
  photons and 120, *122*
  positive-energy 77
  and positrons 80, 108, *122*, 131
  and radioactive decay 132, 133
  self-energy of 73, 74, 108
  self-interaction 74, 75, 76, 129
  'spin-zero' 112
  unaffected by the strong force 191
  and uncertainty 130
  virtual 131

and virtual photons 129, *129*, 199
wave behaviour 35, 36
and weak interactions 132, *133*
zero radius of 74
electrostatics 72
electroweak theory of particle physics *133*, 134
Emde 113
Encyclopaedia Britannica 1, 2, 15, 170, 217
energy
conservation of 17
kinetic 17, 190
potential 17
Engineering and Science 145, 150, 151
Esalen Institute, Big Sur 245, 246, 249
ETH (Federal Institute of Technology), Zurich 142
Ethical Culture Institute, New York 6
Euclid 163
Euclidian geometry 13
evolution xv, 227
'experiment with two holes' 37–9, *38*, 49, 121, 123, 276–7, *276*, 278

Far Rockaway, Queens County, New York 1, 2, 6, 7, 12, 43, 46, 48, 59, 99, 246, 257
Far Rockaway High School 11, 15
Faust (Goethe) 53–4, 55
Fermi, Enrico 65, 92, 156
Fermilab 135
fermions 156, 157, 191
Feynman, Anne (F.'s paternal grandmother) 4
Feynman, Arline (née Greenbaum; F.'s first wife) 48, 63, 68, 114, 139, 145, 146, 248
corresponds with F. from hospital 68, 94
death 99, 102, 103–4, 283
encourages F. not to conform 66
engaged to F. 48
illness and hospitalization 65, 66, 67, 87, 93, 94, 96–7, 99
marries F. 65, 67
start of relationship with F. 21–2, 23
train journey to Los Alamos 93–4
Feynman, Carl (F.'s son) 152, 178, 182, 217, 222
Feynman diagrams 116, 126–9, *126*, *128*, *129*, 131, 133, 134, 136, 158, 197, 224, 262, 270, 283, 284
Feynman, Gweneth (née Howarth; F.'s third wife) 174, 184, 242, 254, 257
adventurousness 149, 182, 204
and the *Challenger* inquiry 229, 239
childhood 149–50
death 258
F. meets 149, 150, 168
and F.'s death 257
as F.'s housekeeper 151–2
grave in Altadena 282
illness 218, 258
in Japan 252
marries F. 152, 153, 168
in Stockholm 181
visits Tannu Tuva 258
Feynman, Henry Phillips (F.'s brother) 7
Feynman, Jakob (F.'s paternal grandfather) 4
Feynman, Joan (F.'s sister) 198
advises F. on understanding the Lee paper 162, 164
and Arline 21, 48
and the aurora borealis 9, 10, 212
birth 7
career in science 8, 9, 10, 162, 250
education 57
on F.'s anecdotes xvi
as F.'s 'first student' 8–9
on F.'s messages when dying 257–8
on her father's observation of his own illness 243
influence on F. 8
IQ test 19
and Lucille as storyteller 6–7
moves to be near F. 250
Feynman, Lucille (née Phillips; F.'s mother)
and Arline 48
birth 6
death of second son 7
discourages Joan from being a scientist 9

education 6
family background 5
and F.'s birth 1, 6
and F.'s education 22
influence on F. 6–7, 13
meets and marries Melville 6
opposes F.'s marriage to Arline 67
personality 6, 56, 209
provision made for 57
as a storyteller 6–7, 209
teaches F. laughter and human compassion 3, 6
Feynman, Mary Louise (née Bell; F.'s second wife) 148, 153
as an art history student at Cornell 142
divorce 147, 149, 153, 160, 161
F. proposes by letter 145–6
honeymoon 146, 205, 206
in Japan 148
marries F. 143, 146
tries to cut F.'s social contact with scientists 146
Feynman, Melville (F.'s father)
and anti-semitism 57, 62
Arline paints with 48
corresponds with F. in code 94
death 103
encourages F.'s interest in science 1–2, 3, 6, 8, 48
family background 4
and F.'s education 22, 23, 57
health problems 57, 243, 249
and Joan's interest in science 8
marries Lucille 6
opposes F.'s marriage to Arline 67
relationship with F. 2, 182
as a storyteller 2, 3–4, 144
teaches F. the value of money 20
in the uniform business 5, 7
Feynman, Michelle (F.'s daughter) 152, 182, 195, 204–5, 208, 217, 262–3
Feynman, Richard Phillips
ability to explain physics 73
appearance 47, 138
and Arline's death 99, 102, 103, 104, 283
and art 55, 183–4, 208, 219, 242
becomes the leading physicist of his generation 118
birth (11 May 1918) 1, 6
at Caltech *see under* California Institute of Technology
carries the quantum mechanics theory to its greatest fruition 43
and *Challenger see* Rogers Commission
as the 'Chief' 244
childhood education 9, 10–16, 19, 22, 82
completion of his PhD 64, 65, 67, 87, 88, 92
and computers 222–7
at Cornell *see under* Cornell University
and cosmic rays 70–71
death (15 February 1988) xv, 257–8
divorce from Mary Lou 147, 149, 153, 160, 161
and Dresden Codex 205–6, 207
drumming 6, 55, 62, 116, 147, 182, 208–11, 218, 219–20, 244, 246, 247
Dyson on 112
father encourages his interest in science 1–2, 3, 6, 8
and father's death 103–4
first published 56
first scientific contacts 11
first trip to Europe 142
health problems 203, 218–19, 227, 228, 239, 243, 244, 245, 248–50, 254, 257, 258
his children 152
indifferent to publication 163
integrity 158–9, 217, 248, 279
Joan influences 8
'ladykiller' reputation 139, 152
last public appearance 255
and Manhattan Project *see* Manhattan Project
marries Arline 65, 67, 92
marries Gweneth 152, 153, 168
marries Mary Lou 143, 145
Messenger Lectures 178–9, 210
Nobel lecture (Stockholm, 1965) 73, 74–5, 80, 83, 85, 117, 136, 180–81

Nobel Prize for Physics 1, 104, 108, 137, 152, 179–82, 185, 186, 284
personality xiii, xiv, xvi, 101, 138, 217, 282–3, 284
Pocono Conference lecture (1948) 110, 115
and Proctor Fellowship 63
receives Albert Einstein Award 148
refuses honorary degrees 183
research assistantship at Princeton 57–8
in Rio 140–41, 143–6, 165
safecracking 6, 94, 209, 212
as a showman 246–7, 255
starts at MIT 29, 43, 44, 45–8
story-telling xvi, 3, 4, 104n, 116, 182, 220, 248, 256, 258–9
as a teacher xiv, xv, 97, 103, 104, 169, 171, 175, 176, 178, 204, 248
teaches Joan about science 8–9
trip to Albuquerque with Dyson 114–15, 149, 247–8
his van 262–3, 283–4
visits Japan 148, 252
works: The Art of Richard P. Feynman 242; *The Character of Physical Law* 178, 210; *The Feynman Lectures on Gravitation* (Feynman, Morinigo and Wagner) 261, 262, 268, 270; *The Feynman Lectures on Physics* xv, 11, 174–7, 185, 208, 251, 278; 'Forces in Molecules' 73; 'Los Alamos from Below' 92; *QED: The Strange Theory of Light and Matter* 121, 125, 136, 177–8, 247, 250; 'The Quantum Mechanical View of Reality' (lectures) 246; *Six Easy Pieces* 177; *Surely You're Joking, Mr Feynman!* (with Leighton) 20, 46–7, 53, 61, 65, 102, 103, 104n, 139, 142, 144, 161, 164, 179, 206, 210–11, 244, 247, 250; 'There's Plenty of Room at the Bottom' (talk) 170; *What Do You Care What Other People Think?* (with Leighton) 13, 21, 66, 67, 96, 97, 113, 236, 239, 247
Field, Richard 201–2
field theory 26–7, 74
Finnegans Wake (Joyce) 194
Fokker, Adriaan 79
Ford Foundation 174
Fowler, Willy 152–3, 163, 164, 214, 217, 268
Fox, Geoffrey 202
Frankfort Arsenal, Philadelphia 63
friction 48, 56
Friedman, Jerome 197, 199
Frogs, The (Aristophanes) 180
From Eros to Gaia (Dyson) 178, 242
Fuchs, Klaus 99
fundamental particles 271–2
FVH technique 169

gamma rays *122*
Gamow, George 266
Gast, Harold 22
Gell-Mann, Murray 215, 261
at Caltech 160, 165, 190, 193, 211, 282
classification of elementary particles 191–2, 195
and current algebra 198
and the eightfold way 191
Nobel Prize (1969) 192, 195
and quarks 193–4, 195
and strong interaction 189
and weak interaction 165–8, 189, 190, 194
General Electric Company 106
General Theory of Relativity *see under* relativity
Geneva 149, 150, 151
geometry 13, 15–16
'ghost' fields 269–70
Gianonni 184, 185
Gleick, James 151, 158
gluons 135, 136, 199, 201, *202*
Goethe, Johann Wolfgang von 53, 54
Goldberg-Ophir, Haim 192, 193
Goodstein, David 175, 176, 185–6, 189, 219, 255, 279
Graham, William 228, 229, 233, 236
Grand Unified Theory (GUT) 136
gravitational fields 26
gravitational radiation 164

gravitons 164, 261–2, 269, 272
gravity 30, 80, 118, 120, 134, 271
  F.'s postgraduate lectures 178
  F.'s work on 163–4, 171, 173, 178, 185, 189, 192, 260–63, 266, 268, 269, 270, 272
  and the inflation theory 265–6
  lack of a completely satisfactory quantum theory of 263
  Newtonian 25, 266, 270
  and the perturbation technique 261
  and superstring theory 271, 272
  and a Theory of Everything 136
  universal law of gravitation 25
Gribbin, John xv, 176, 268, 281
GUT *see* Grand Unified Theory
Guth, Alan 267–8
Guys and Dolls 209–10

hadrons 191, 192, 194, 195, 196, 199, 200
hallucinogens 244–5
Hamilton, William 52
Hamiltonian method 52, 53, 82, 83, 89, 108, 109
Hartle, James 260, 261, 265, 269
Harvard University 105
Hatfield, Brian 261, 262–3
Hawking, Stephen xv
Heisenberg, Werner 36, 71, 88, 130
Heisenberg's Uncertainty Principle 130
Hellwarth, Robert 168, 169
Hibbs, Al 229, 238
Hillis, Danny 222, 223, 225, 227, 258
Hiroshima 100
Hitler, Adolf 63, 65
Hoyle, Fred 163, 268, 269
Hughes Research Laboratories, Malibu (Hughes Aircraft Company) 169, 178, 228, 244, 260
Hunting of the Snark, The (Carroll) 111
Huntington Medical Library 248–9
hydrogen atoms
  Bethe calculates the energy for an electron in 107
  and microwaves 105
  and path integral approach 215–16
hypnosis 62

I Asimov (Asimov) 20
Iben, Icko 268–9
IBM 95, 249
inertia 3, 75, 76, 272, 273
infinities 106, 107, 108, 127, 134, 272
inflation theory 265–6, 268
Institute for Advanced Study, Princeton 81, 103, 109, 111, 113, 160, 214
Institute for Theoretical Physics, Berlin 215
interference fringes 37
interference pattern 37, 38, *38*, 39, 123, 277
intermediate vector bosons *see under* bosons
Introduction to Theoretical Physics (Slater) 49
inverse beta decay 59
IQ tests 19–20
Israel Atomic Energy Commission 192

Jahnke 113
Jehle, Herbert 83, 84, 87, 108
Jet Propulsion Laboratory, Pasadena (JPL) 8, 204, 229, 230, 233, 243, 250, 283
Johnson Space Center 233
Jordan, Pascual 266
Joyce, James 194

Kac, Mark 20
kaon 162
Kapitsa, Andrei 254
Kapitsa, Pyotr 254
Karc, Marc 215
Kellogg Radiation Laboratory, Caltech 144, 268
Kendall, Henry 197, 199
Kennedy Space Center 233
Kleinert, Hagen 215, 216
Kramers, Hendrik 107
Kutyna, Air Force General Donald 228, 230–37
Kyzyl, Tannu Tuva 218, 241, 258

La Cañada High School, Los Angeles 255

Lagrange, Joseph Louis 52
Lagrangian approach 52–3, 82–3, 88, 164, 270
'Lagrangian in quantum mechanics, The' (Dirac) 83–5
Lamb, Willis 105
Lamb shift 105–9, 112, 118, 120, 131, 144
Lamin, Vladimir 258
Lamy, New Mexico 93
Landau, Lev 159, 214
Langevin, Paul 35–6
Las Vegas 142, 149, 168, 183, 245–6
lasers 168, 169
laws of motion *see under* motion
Lee, Tsung Dao ('T.D.') 160, 161, 163, 164
Leighton, Phoebe (née Kwan) 253, 258
Leighton, Ralph 3, 152, 182, 208–9, 210, 217–18, 256, 282
  describes F. as a 'shaman of physics' xiv
  at the Esalen Institute 245, 246
  on F.'s storytelling xvi, 209
  and F.'s van 283, 284
  marries Phoebe 253
  as a musician 208, 209, 210, 219, 244, 246, 247
  role as F.'s scribe 246, 247
  in Sweden 252
  takes F. to Las Vegas 245–6
  and Tannu Tuva 217–18, 241–2, 243, 250–51, 258
Leighton, Robert 174
  works with F. on the *Lectures* 208
leptons 132, 133, 134, 191, 271
LeSur, William 11
Lewine, Frances (F.'s cousin) 7, 257
Lewine, Pearl (née Phillips; F.'s aunt) 7
Lewine, Ralph (F.'s uncle) 7
Lewine, Robert (F.'s cousin) 7, 13
Lewis, Gilbert 33
Library of Congress 170–71, 207
light
  and Maxwell's equations 26
  and the quantum revolution 25, 30
  and relativity theory 25, 33
  speed of 25, 28, 29, 49, 74, 77, 190
  in terms of particles 33, 36
  'travelling in straight lines' theory refuted 124–6, *124*, *125*
  as a wave 26, 33, 36, 37, *37*
Lilly, John 244
Linde, Andrei 268
lines of force 26, *27*
liquid helium 144, 155, 159, 161, 213
liquid helium I 156
liquid helium II 156, 254
Lopes, Leite 143–4
Lorentz invariant 127
Los Alamos National Laboratory 170

MacDonald (of Thiokel) 234–5
Mach's Principle 272–3, 278
McLellan, William 170
magnetic poles 26, *27*
magnetism 25, 143, 173
Manhattan Project 91–100, 105
  Bethe and xv, 94, 95, 96, 100
  F. as a group leader (Theoretical Computations Group) 1, 94–5, 97, 222, 223, 226
  F. starts work (1942) 89
  first nuclear explosion on Earth (Trinity test) 100
  lack of safety precautions 98–9
  nears completion 99
  uranium separation 64, 65, 97
  Wilson tells F. of 63–4
'many worlds' idea 264–5
Marshak, Robert 165, 167, 168
masers 168, 169
Massachusetts Institute of Technology (MIT) 43, 45–56, 64, 68, 70–73, 102, 139, 197
  Artificial Intelligence Lab 225
  Carl studies at 222
  F. applies to 22
  F. changes courses 48–9
  F. gains a scholarship at 23
  F. graduates 56
  F. starts at 29, 43, 44, 45–8
  flexibility of 49
  and Jewish students 45
  Phi Beta Delta 45, 46–8
  Sigma Alpha Mu 46

mathematics, and physics 18, 178
Mautner, Alix 246
Mautner, Leonard 11, 13, 246
Maxwell, James Clerk 25, 26, 173, 177
Maxwell's equations 25–8, 31, 75, 77, 78, 79, 81, 159, 166, 261, 274
Mehra, Jagdish 13, 139, 174, 182, 214, 256, 257
Mendeleyev, Dmitri 191–2
Mensa 20
Meselson, Matt 147, 169
mesons 143, 191, 193, 200, 201, 252
Messenger Lectures 178–9, 210
Metallurgical Laboratory, Chicago 92
Michigan State University 146
microwaves, and hydrogen atoms 105
Millikan, Robert 33
Minsk, Byelorussia 4
Minsky, Marvin 222, 223
Mir publishing house 251
Moore, Lillian 14
Morette, Cecile 116, 141
Morinigo, Fernando 261
Morrison, Philip 92
Morse, Philip 51, 56–7, 71–2
Morton, Dr 245, 253, 257
Most of the Good Stuff (Brown and Rigden, eds.) 197, 213, 227, 250, 255, 279
motion, laws of 28, 52, 75, 82
Mount Wilson Observatory, San Gabriel Mountains 147
Mountain View Cemetery, Altadena 282

Nagasaki 100
nanotechnology 170
NASA (National Aeronautics and Space Administration) 228, 231–4, 236, 238, 239
National Academy of Sciences 104, 118, 148–9
Natural History Museum, Los Angeles 252–3
Natural Selection 227
nature
  F.'s attitude towards xiv, 272, 279
  F.'s father introduces him to 2
  F.'s father's insight 3
  and gravity 120
  and parity conservation 160
  and QED 120
  and the speed of electromagnetic wave movement 25–6
Nature xv, 267
Ne'eman, Yuval 191, 192, 193
negative roots 40–41
Neher, Victor 174
Neugebauer, Otto 207
Neumann, John von 81
neutrino
  and beta decay 190
  discovery of 29–30
  equation of 166–7
  and weak interactions 132, *133*
neutron decay 135, 162, 164
neutron stars 93
neutrons
  affected by the strong force 191
  beta decay of 165
  electrically neutral 29
  electrons bouncing off 117
  identified 29
  and nuclear fission 91
  quarks and 134, 135
  and radioactive decay 132
  slowing down 97–8
  and weak interactions 132, *133*
Newman, Tom 170
Newton, Sir Isaac 23, 25, 28, 52, 75, 82, 88, 177, 266, 269, 270
No Ordinary Genius: The Illustrated Richard Feynman (ed. Sykes) 204
Nobel, Alfred 161
Nobel Committee 42, 179, 192
nuclear fission 91
Nuclear Research Laboratory, Cornell 103
nucleus
  and nuclear fission 91
  structure of 29, 144
Nuovo Cimento, Il 192

O rings 229–30, 252
Oak Ridge, Tennessee 97, 98, 99
'Ofey' 184

Oldstone Conference (1949) 118
Oldstone-on-the-Hudson, Peekskill, New York 118
'On the Silk Road' exhibition 251, 252, 258
Onnes, Kamerlingh 155–6
Oppenheimer, Robert 6, 96, 115, 118, 161
  death 93
  as director of the Institute for Advanced Study 111, 113, 214
  and the National Academy of Sciences 104
  as scientific head of the Manhattan Project 93, 97, 98
  and students 212
  tries to lure F. to Berkeley 101
optics 51, 123, 124, 158
orbital 39

Pachos, Emmanuel 199
parallel processing 222–3, 225–6, 227
parallel realities 58
parity conservation/violation 160, 161, 162
particle accelerators 61, 135, 190, 193, 194, 195, 197
particles 33, 88
  and antiparticles 127, 133
  interactions 43, 164–5, 198, 261
  virtual 164–5
partons 196–7, 198, 199, 202, 205, 208
Patchogue, Long Island 4
path integral approach ('sum over histories' approach) 215–16, 265, 270, 273, 277, 278
  see also under quantum electrodynamics; quantum mechanics
Pauli, Wolfgang 81
Payne-Gaposhkin, Cecilia 9
Pearl Harbor 64
Periodic Table 191–2
perturbation technique 261, 269
Phillips, Henry (F.'s maternal grandfather) 5, 7
Phillips, Johanna (née Helinsky; F.'s maternal grandmother)
  marries Henry Phillips 5
  works with her father in New York 5
photoelectric effect 33
photons
  and electrons 120, *122*
  and experiment with two holes 38, *38*, 49
  and gravitons 164
  named 33
  scattering by an electric field 116
  scattering by other photons 116
  virtual 129–31, *129*, 136, 199
Physical Review 56, 61, 71, 73, 115, 127, 166, 167, 199
physics
  atomic 173
  conservation of energy 17
  experimental 51
  Feynman family's love of 9
  F.'s father's insight 3
  F.'s lectures on introductory 174–8
  F.'s love of xv, xvi, 104, 145, 250, 284
  F.'s most important contribution to 176–7
  as fun xiv, 104, 145, 250
  high energy 215
  and mathematics 18, 178
  particle 127–8, 166, 197, 199
  solid state 10, 162, 215
  theoretical nuclear 51–2
Physics Letters 194
Physikalische Zeitschrift der Sowjetunion 83
Pines, David 158
pions 160
Planck, Max 23, 32, 33, 34
Planck length 270–71
Planck's constant 32, 35, 271
Playa de la Mision, Baja California 182
plutonium, critical mass 98
plutonium-239 91
Pocono Conference (1948) 108, 110, 111, 112, 115, 118
positrons
  and electrons 80, 108, *122*, 131
  named 42
  positively charged 42, 71
  virtual 131

Preskill, John 268
Princeton 93
Princeton University 9, 56–68, 102, 103, 139, 266
  bicentennial celebration (1946) 84
  cyclotron 61
  F. leaves (1943) 93
  F. rejects honorary degree 183
  F. starts research at 73, 76
  first-class physics school 61
  imitates Oxford and Cambridge 61
  and Jewish students 56–7
  offers F. a research assistantship 57–8
Principle of Least Action 16–18, *19*, 16–18, *19*, 52, 74n., 79, 82, 89, 125, 177, 225
Principle of Least Time 18, *19*
Principles of Quantum Mechanics, The (Dirac) 43–4
protons
  affected by the strong force 191
  magnetic moment of 136
  positively charged 29, 42
  quarks and 134, 135
  and radioactive decay 132
  role of the 29
  and weak interactions 132, *133*
punctuated equilibrium 227

QCD *see* quantum chromodynamics
QED *see* quantum electrodynamics
QED and the Men Who Made It (Schweber) 111
quanta, Planck and 32
quantum chromodynamics (QCD) 135–6, 201–3, *202*, 227, 254, 256, 257
quantum electrodynamics (QED) 212, 262, 281
  Bethe's discovery 106–8
  Dyson's paper 115
  F. describes his path to 180–81
  F.'s approach 108–11, 113–18, 121, 137, 141, 144, 155, 158, 159, 166, 168, 181, 186, 202, 224, 225
  and infinities 106, 107, 108, 127, 134
  and QCD 136, *202*, 292
  and Rabi's discovery 105
  success of 120–21, 131, 134
quantum gravity 164, 165
quantum of length 271
quantum mechanics 25, 29, 40, 72, 73, 82, 175, 177, 263, 264
  'central mystery' of 37, *38*
  and classical mechanics 89
  F. carries the theory to its greatest fruition 43
  at MIT 51
  and modern understanding of chemistry 40
  path integral ('sum over histories') approach 85–9, 106, 123–4
  in the presence of gravity 273
  spacetime approach 85, *86*, 106
  transactional interpretation of 278, 281
quantum physics 12, 30, 173, 176
  becomes F.'s vocation 23
  F.'s greatest work in xv, 60
  and the Lagrangian approach 83, 84, 85
  wave-particle duality as a key ingredient 36
quantum probability 39, 122–3, 125, 264, 274, 275, 277, 278
quantum revolution 30
quarks 134–5, 136, 193–5, 199–201, *202*, 226, 271
quasars 163, 268

Rabi, Isidor 16, 105
radiation
  electromagnetic 190, 281
  gravitational 164
  Wheeler–Feynman theory (absorber theory) 73–82, *78*, 121, 273
radiation resistance 75, 76, 77, 79, 272, 273
'Radiation Theories of Tomonaga, Schwinger and Feynman, The' (Dyson) 115, 116
radio 12, 16, 75–6
Radio Moscow 241, 242
radio waves 28
radioactive decay 132–3

Raramuri people 205
relativistic pancakes 198, 200
relativity 30, 107, 173
  built into F.'s version of quantum theory 108
  F. teaches himself in high school 19
  Feynman family's interest in 10
  General Theory of (1916) 23, 29n, 50, 75, 120, 121, 163, 261, 266, 268, 270, 272, 282
  Special Theory of (1905) 23, 25, 28, 29, 40, 112, 131
renormalization 107, 127–8, 134, 136, 262, 270
Reserve Officer Training Corps (ROTC) 50, 53, 55
Retherford, Robert 105
Reviews of Modern Physics 82, 87, 109
Ride, Sally 228
Rigden, John 255, 279
Rio de Janeiro 140–41, 143–6, 165
Robertson, H. P. 63
'Rochester' conference, Switzerland (1958) 168n.
Rochester, New York conference (1956) 160
Rochester, New York conference (1957) 162, 164, 167
Rogers, William 228, 230, 232, 233, 235
Rogers Commission 68, 228–39, 252, 253
Russell, Henry Norris 81
Rutherford, Ernest 34
Rutishauser, Tom 208, 209

S (scalar) interactions 165, 167
Salam, Abdus 132
Sands, Matthew 151, 152, 173–4, 176
scale invariance 198
Schenectady, New York 106, 108
Schrieffer, Robert 158–9
Schrödinger, Erwin 36, 138, 263, 264
Schrödinger equation 36–7, 84, 88, 89, 108, 122, 216, 274, 275, 276
Schrödinger's cat paradox 263–4, 277, 278
Schweber, Silvan 111, 113
Schwinger, Julian 171, 198
  American Physical Society meeting (1948) 109
  at Ann Arbor 114, 115
  calculates the magnetic moment of the electron 109
  compares notes with F. 110–11
  complicated method 109–10, 111, 127, 128
  Dyson and 115
  Pocono Conference 110, 118
  shares Nobel prize with F. and Tomonaga 179, 180
  at the Shelter Island Conference 105
  version of the Lamb shift calculation 109
science, key to 179
Scientific American 246–7
Scituate 48
Segre, Emilio 97
sensory deprivation 244
Shaw, Christopher (F.'s nephew) 152
Shaw, Jacqueline (née Howarth; F.'s sister-in-law) 149, 150, 152
Shelter Island Conference 105, 106, 107, 118
Sherman, Richard 215
Shermer, Michael 282, 283, 284
Shockley, William 159
Skeptics Society 282
Slater, John 49, 50, 56, 72
Slotnick, Murray 117, 163
Smyth, H. D. 56
solid geometry 15–16
South Pacific 243–4
Soviet Academy of Sciences 252–3, 254, 255, 258
Soviet Ministry of Culture 253
Sovinfilm 253
space, 'across the page' 85, *86*
spacetime 276
  diagrams 85, *86*, *122*, 126–7, *126*
  F. visualizes quantum processes in 114
  as flat 267
  in four dimensions 262
Special Theory of Relativity *see under* relativity

spectra, and Schrödinger's wave equation 37
standing wave 35
Stanford Linear Accelerator Center (SLAC) 197, 198, 199
Stanford University 170, 197, 198
Steady State hypothesis 266
stellar structure 268–9
Stratton, Julius 51
strong force/interaction 29, 30, 131–2, 134, 135, 189, 191, 199, 200, 201
subatomic particles
  discovery of 42
Sudarshan, George 165, 167, 168
'sum over histories' approach *see* path integral approach
superconductivity 156, 158, 159, 215
superfluidity 156–9, 160, 168, 171, 186, 254
supermassive stars 163, 268, 269
superposition of states 264, 277
superstring theory 271–2, 273
Sykes, Christopher 256, 257

T (tensor) interactions 165, 167
Tale of Two Cities, A (Dickens) 170
Tannu Tuva 217–18, 219, 241–3, 245, 250–56, 258
tau 160, 161, 162
Taylor, Richard 197, 199
Theory of Everything (TOE) 136
theta 160, 161, 162
Thinking Machines Corporation 226, 229, 241, 242
Thiokel 234–5, 238
Thomas, Dylan 281
Thompson, Eric 206
Thompson, J. E. 15
Thompson, S. P. 15
Thomson, George Paget 36
Thomson, J. J. 29, 36
Thorne, Kip 213, 268, 282
Three Quarks 209
time, 'up the page' 85, *86*
Tiomno, Jaime 140
Titan rocket 232
TOE *see* Theory of Everything
Tomonaga, Shin'ichiro 111, 127, 179, 180, 278, 281
transactional interpretation of quantum mechanics 278, 281
transistor effect 159
Trinity test 100
Tryon, Edward 267
Tuck, Helen 211, 212, 219, 248, 282
Tukey, John 62
Tuva or Bust! (Leighton) 251
Tuvan Research Institute of Language, Literature and History 242

Ukonu 208, 210
ultraviolet catastrophe 31–2
uncertainty 130–31
United Nations (UN) 149
Universal Law of Gravitation 25
Universal Studios 210
Universe 263
  the age of the 147
  Big Bang theory 266
  big crunch 267
  expansion of 267
  as flat 273
  inflation theory of its origin 265–6, 268
  Steady State hypothesis 266
  tiny seed of the 267
  'the total energy of the universe is zero' hypothesis 266–7
University of California, Berkeley, California 92, 101–2
University of California, Los Angeles (UCLA) 143, 207, 246
University of Chicago 148, 183, 185
University College of Los Angeles Medical Center 243, 253, 254, 257
University of London (England) 191
University of Michigan, Ann Arbor 114, 115, 193
University of North Carolina, Chapel Hill (conference on the role of gravitation in physics, January 1957) 164
University of Rochester 165
University of Sussex 176

University of Tokyo 252
University of Washington, Seattle 274
University of Wisconsin 101
uranium-235
  critical mass 97, 98
  nature of 91
  separating from uranium-238 64, 91–2, 97
uranium-238
  nature of 91
  separating uranium-235 from 64, 91–2, 97
US State Department 143

V (vector) interactions 165, 166, 167
Vainshtein, Sevyan 251
Vallarta, Manuel 70, 71
variational principle 216
Vernon, Frank 168
Vinita, Oklahoma 114, 149

Wagner, William 261
Warsaw conference (July 1962) 164
Watson, James 185, 186, 195
wave–particle duality 36, 39
waves 88, 121
  advanced 77, 78, 121, 273–4, 275, 277, 278
  confirmation 275, 276, *276*, 277
  offer 275, *276*, 277
  out of phase *30*, 78, 121
  in phase *30*, 78, 121
  retarded 77, 78, 79, 121, 273–4, 275, 277
wave concepts 49
wave equations 37
weak force/interaction 30, 132, 133–4, *133*, 161, 165
  theory of weak interaction 159–60, 162–3, 164, 165–8, 171, 186, 189, 190, 194, 195
Weinberg, Steven 116, 132
Weiner, Charles 205, 256
Weisskopf, Viktor 181
Welton, Ted 50–51, 52, 55
Wheeler, John 94, 96, 261, 272
  and Arline's illness 65
  becomes F.'s thesis adviser 56
  examines F.'s thesis 65
  F. works at Wheeler's house 62
  and F.'s entry to Princeton 56
  Melville discusses F.'s career with 62
  and quantum mechanics 73
  research with F. 64
  start of friendship with F. 58–9
  tells Einstein of F.'s path integral approach 88–9
  and the Wheeler–Feynman theory of radiation 76–80, *78*
  works on the world's first nuclear reactor 65
Wheeler–Feynman theory of radiation (absorber theory) 73–81, *78*, 121, 273, 275
Wigner, Eugene 65, 81, 101
Wilson, Robert 64–5, 91, 92, 103
world lines 85
World's Fair, Chicago 16
Wu, Chien Shiung 161, 165

Yang, Chen Ning ('Frank') 160, 161, 163
Yukawa, Hideki 252

zero shift 106
Zorthian, Jirayr ('Jerry') 183, 184
Zurich 142
Zweig, George 192–5